Computational Methods in Applied Sciences

Volume 45

Series editor

E. Oñate
CIMNE
Edificio C-1, Campus Norte UPC
Gran Capitán, s/n
08034 Barcelona, Spain
onate@cimne.upc.edu

More information about this series at http://www.springer.com/series/6899

Pedro Diez · Pekka Neittaanmäki
Jacques Periaux · Tero Tuovinen
Olli Bräysy
Editors

Computational Methods and Models for Transport

New Challenges for the Greening of Transport Systems

Springer

Editors
Pedro Diez
CIMNE, International Center for Numerical Methods in Engineering
Barcelona
Spain

Pekka Neittaanmäki
Department of Mathematical Information Technology
University of Jyväskylä
Jyväskylä
Finland

Jacques Periaux
Department of Mathematical Information Technology
University of Jyväskylä
Jyväskylä
Finland

Tero Tuovinen
Department of Mathematical Information Technology
University of Jyväskylä
Jyväskylä
Finland

Olli Bräysy
Department of Mathematical Information Technology
University of Jyväskylä
Jyväskylä
Finland

ISSN 1871-3033
Computational Methods in Applied Sciences
ISBN 978-3-319-54489-2 ISBN 978-3-319-54490-8 (eBook)
DOI 10.1007/978-3-319-54490-8

Library of Congress Control Number: 2017937298

Printed on acid-free paper

This Springer imprint is published by Springer Nature
The registered company is Springer International Publishing AG
The registered company address is: Gewerbestrasse 11, 6330 Cham, Switzerland

Foreword

We are seeing very exciting times for people working in transport and logistics. On the one side, service providers and regulators face incredible challenges by globalization, which extends the supply and distribution networks across all continents, the sustainability issues related to transport, the tremendous progress in vehicular technologies, and the evolution of the demand toward more and more personalized services. On the other side, considerable research efforts were performed in the last decades in optimization and data science. Such efforts start to bring new and powerful computational methodologies that are capable to cope with the complexity of decision-making in such a rapidly evolving arena.

In the era of Big Data, modern decision-makers in transport and logistics must be able to deal with enormous amounts of information and a variety of heterogeneous data sources: from the traditional enterprise information systems to myriads of mobile devices and autonomous sensors which are available in the Internet of Things. Furthermore, new vehicles with electric or hybrid engines, as well as the autonomous vehicles which are already diffused in industrial plants and will soon access the road network introduce totally new problems and opportunities. Decisions have largely widened their scope moving toward large-scale and integrated time horizons in which several components of the supply chain or transport planning are jointly considered, allowing for increased efficiency and robustness of the solutions. These continuously changing scenarios thus require innovative paradigms both to predict short-term events and long-term trends and to draw optimized decision patterns taking into account data variability and richness, as well as managing sustainability objectives on profitability, environment and peoples (which can be conflicting).

The attention to environmental sustainability in the recent years has attracted a huge attention of the researchers and of the industry, motivated by the fast growth in Greenhouse Gas (GHG) emissions produced by transport modes over the last decades. To mitigate this trend, several technologies or behavior-based solutions have been developed. On the technological side, carbon intensity can be lowered through the introduction of biofuels and alternative energy carriers (electricity and hydrogen). Additionally, the energy intensity of transport and mobility can be lowered by developing more energy-efficient vehicles. The diffusion of electric vehicles is today modest, mainly due to the higher cost and limits in travel autonomy with respect to the traditional fuel vehicles. In addition, plug-in electric vehicles require access to charging stations which, in particular for the fast charging ones still require considerable public investments. However, the progress in battery technologies, pushed not only by the automotive sector, and the fast success of hybrid vehicles, now competitive even costwise with traditional ones, open a bright future for alternative vehicles not only in people transportation but also for freight delivery particularly for the last mile operations in urban areas which are increasingly inaccessible to combustion vehicles. From the optimization and management point of view, electric and hybrid vehicles present new and complex challenges for which the study is still at an early stage. The new problems originate from the limited autonomy, the use of intermediate charging stations and the need to model accurately the consumption and charging rates of the batteries. New challenges also originate from the interaction between electric/hybrid and traditional combustion vehicles in the supply chain where environmentally friendly may be employed to reach customers in multi echelon distribution networks.

On the behavioral side, transport GHG emissions can be reduced through policy choices that help decision-makers choosing transportation options that will lower GHG emissions. The modal share of passenger mobility and freight transport can be shifted from carbon-intensive options (air, passenger cars, and trucks) to less carbon-intensive ones (public transport and nonmotorized modes for passengers; rail and inland waterways for freight). As to passenger mobility new use models are now a reality, also thanks to many apps that favor crowd participation to new services such as car-sharing and pooling. Car-sharing systems are increasingly employing environmentally friendly electric vehicles. As in freight transportation applications, the design and management of Ecar-sharing systems poses several additional challenges with respect to those based on traditional combustion vehicles, related with the limited autonomy of the batteries of electric vehicles.

The impact and cost of transport can be decreased by a more efficient organization and allocation of transport movements and its efficiency can be boosted by the introduction of radically new design of distribution networks which step away from intense use of long distance hauling, toward the coordinated sequencing of shorter shipping according to the Physical Internet paradigm. This is a tremendous change in the structure of future freight distribution and transportation which need substantial investments in shared infrastructures and new and handling and packaging technologies. However, the cost-effectiveness of the new systems mostly relies on three main innovations in the freight transportation. The first is the development of standardized transport containers for truck and train shipping of smaller size with respect to those used in maritime transportation. Second is the consequent massive introduction of automated handling of the containers to reduce the unloading, transfer, and loading operations at terminals. Finally, it is strongly required the design of fast and accurate optimization algorithms that synchronize and plan all transportation activities. Such new tools must incorporate Big Data predictive methods capable of anticipating the dynamic evolution of traffic conditions and shipment progress. With the well-advanced study and prototyping of all these technologies within several research projects across the globe there are encouraging signals that we are all about to see a revolution in goods transportation that will have an even larger impact than that was achieved with the introduction of shipping containers in 1956.

The CM3 Conference on Computational Multi physics, Multi Scales and Multi Big Data in Transport Modeling, Simulation and Optimization was a perfect environment to set the state of the art in many aspects of the new trends in transportation and logistics research and practice as it is witnessed by the content of this excellent volume. Most of the current hot topics in the research and practice arenas are represented in the program from the features of Internet of Things to the coordination of unmanned vehicles. Several specific methodological problems related with green logistics and transportation are also examined in detail as, for example, environmentally friendly route allocation or fleet size and mix problems.

November 2016

Daniele Vigo
DEI "Guglielmo Marconi"
University of Bologna
Bologna, Italy

Preface

The CM3 2015 Conference (Computational Multi physics, Multi Scales and Multi Big Data in Transport Modeling, Simulation and Optimization) is the first of a series on European Conferences on Green Challenges in Transport which took place at Jyväskylä, Finland on May 25–27, 2015.

Its aim was bringing together academic and small and large industries experts in the fields of Computational Methods and Tools for Transport and its Applications.

This series of Thematic Conferences on Transport was originally launched by the Industry Interest Group (IIG) of ECCOMAS with main objectives are: strengthening the industrial liaison and developing contacts between ECCOMAS and the different DGs of the European Commission.

The conference included: introductory lecture (1), plenary (4) and semi plenary (6) lectures, Special Technology Sessions (2) on Aeronautics and Maritime, Round Tables (3) on New Challenges and Solutions for Greening Transport, The Growing Importance of Intra-Logistics and the Role of Information and Big Data, Parallel Contributed sessions (7).

The Transport disciplines considered during the conference were Aeronautics, Automotive, Logistics, Maritime and Railways, with the Logistics as dominant discipline (1,5 day of the program dedicated to SMEs involved in Logistics).

The content of this volume is organized into three main sections with 15 contributions in the above disciplines classified as follows:

1. Reviews and Perspectives
2. Computational methods
3. Translational Research

The book is addressed to researchers and technologists experts in the fields of "Greening Transport," ranging from innovative computational methods to software tools used for solving challenging industrial and societal problems in Transport.

CM3 2015 was organized by University of Jyväskylä in association with European Commission (EC) DG Research and Innovation, ECCOMAS and the International Center for Numerical Methods in Engineering (CIMNE).

The editors are grateful to Ms. Kati Valpe, Ms. Marja-Leena Rantalainen, from University of Jyväskylä for providing us a professional help in the logistic organization and the setup of the final program of the conference and friendly welcome and help to participants before and along this event. The editors would also like to thank Ms. Jaana Mähönen and Mr. Jarno Kiesiläinen for helping in the technical editing of the book.

Finally, we would like to express our gratitude to Ms. Nathalie Jacobs, Senior Springer Publisher and her staff and to Prof. E. Onate, CIMNE Director and Editor of the Series Lectures Notes in Numerical Methods in Engineering and Sciences for their fair patience in receiving the material of this volume.

Jyväskylä, Finland
November 2016

Pedro Diez
Pekka Neittaanmäki
Jacques Periaux
Tero Tuovinen
Olli Bräysy

Contents

Contributors

Niels Aalund Western Gulf Maritime Association, Houston, TX, USA

Lucio De Abreu Correa Laboratoire MSSMat UMR 8579, CentraleSupelec, CNRS, Châtenay-Malabry, France

Niklas Åkerblom Volvo Car Corporation, Gothenburg, Sweden

Estelle Bongini SNCF, Direction de l'Innovation et de la Recherche, Paris, France

Oliver M.F. Browne ETSIA-UPM (School of Aeronautics), Universidad Politecnica de Madrid, Madrid, Spain

Olli Bräysy Department of Mathematical Information Technology, University of Jyväskylä, Jyväskylä, Finland; VU University Amsterdam, Amsterdam, HV, The Netherlands; Procomp Solutions Oy, Oulu, Finland

Gabriel Bugeda CIMNE, Barcelona, Spain; Universitat Politecnica de Catalunya - BarcelonaTech, Barcelona, Spain

Miguel Chavez School of Aeronautics, Universidad Politecnica de Madrid, Madrid, Spain

Michael Cochez Faculty of Information Technology, University of Jyväskylä, Jyväskylä, Finland; Fraunhofer Institute for Applied Information Technology FIT, Sankt Augustin, Germany; RWTH Aachen University, Aachen, Germany

Mikael Collan Lappeenranta University of Technology, Lappeenranta, Finland

Sofia Costa d'Aguiar SNCF, Direction de l'Innovation et de la Recherche, Paris, France

Regis Cottereau Laboratoire MSSMat UMR 8579, CentraleSupelec, CNRS, Châtenay-Malabry, France

Ralf Deiterding University of Southampton, Southampton, UK

Wout Dullaert Faculty of Economics and Business Administration, VU University Amsterdam, Amsterdam, Netherlands; Institute of Transport and Maritime Management Antwerp, University of Antwerp, Antwerp, Belgium

Baldrik Faure SNCF, Direction de l'Innovation et de la Recherche, Paris, France

William Fitzgibbon Departments of Engineering Technology and Mathematics, University of Houston, Houston, TX, USA

Patrik Flisberg The Forestry Research Institute of Sweden, Uppsala, Sweden

Moritz M. Fragner German Aerospace Center, Göttingen, Germany; Numerische Strömungs- und Gasdynamik BTU Cottbus - Senftenberg, Cottbus, Germany

Mikael Frisk The Forestry Research Institute of Sweden, Uppsala, Sweden

Mario Guajardo Department of Business and Management Science, NHH Norwegian School of Economics, Bergen, Norway

Arild Hoff Molde University College, Molde, Norway

Lars Magnus Hvattum Molde University College, Molde, Norway

Antoine Kalmbach NFleet Oy, Jyväskylä, Finland

Mariia Kozlova School of Business and Management, Lappeenranta University of Technology, Lappeenranta, Finland

Alexander Krylatov Saint-Petersburg State University, Saint-Petersburg, Russia

Tommi Kärkkäinen Department of Mathematical Information Technology, University of Jyväskylä, Jyväskylä, Finland

Pasi Luukka Lappeenranta University of Technology, Lappeenranta, Finland

David Mester Institute of Evolution, Mathematical and Population Genetics Laboratory, University of Haifa, Haifa, Israel

Eugenio Oñate CIMNE, Barcelona, Spain; Universitat Politecnica de Catalunya - BarcelonaTech, Barcelona, Spain

Ghazaleh Panahandeh Volvo Car Corporation, Gothenburg, Sweden

Urooj Pasha Molde University College, Molde, Norway

Jacques Periaux CIMNE, Barcelona, Spain; Faculty of Information Technology, University of Jyväskylä, Jyväskylä, Finland

Jordi Pons-Prats CIMNE, Barcelona, Spain; Faculty of Information Technology, University of Jyväskylä, Jyväskylä, Finland

Pasi P. Porkka Justecon Oy, Helsinki, Finland

Tuukka Puranen NFleet Oy, Jyväskylä, Finland

Jussi Rasku Department of Mathematical Information Technology, University of Jyväskylä, Jyväskylä, Finland

Arian Razmi-Farooji MyOpt Consulting Ltd, Suolahti, Finland; Industrial Engineering and Management, University of Oulu, Oulu, Finland

Mikael Rönnqvist Département de génie mécanique, Université Laval, Québec, QC, Canada

Silvia Sanvido ETSIA-UPM (School of Aeronautics), Universidad Politecnica de Madrid, Madrid, Spain

Vagan Terziyan Faculty of Information Technology, University of Jyväskylä, Jyväskylä, Finland

Tero Tuovinen Faculty of Information Technology, University of Jyväskylä, Jyväskylä, Finland

Eusebio Valero ETSIA-UPM (School of Aeronautics), Universidad Politecnica de Madrid, Madrid, Spain

Charles Voivret SNCF, Direction de l'Innovation et de la Recherche, Paris, France

Dmitriy Volf Saint-Petersburg State University, Saint-Petersburg, Russia

Victor Zakharov Saint-Petersburg State University, Saint-Petersburg, Russia

Francisco Zarate CIMNE, Barcelona, Spain

Part I
Reviews and Perspectives

Chapter 1
The Revolutionary Internet of Things

Arian Razmi-Farooji

Abstract Although Internet of Things (IoT) has got lots of attention especially during recent years, its origin gets back to older times. IoT is all about connecting different entities in systems, equipped with sensors and actuators by means of internet technology. It has various applications in a wide range of industries from healthcare and agriculture to automotive, manufacturing and supply chain. Looking at the huge amount of money, invested by all big players in different industries to provide the necessary technological infrastructure, shows the strategic importance of implementing this game-changing technology. IoT is not a future paradigm; it is all about the present. It has changed the business models, caused new businesses to be introduced and has revolutionized the way that industries perform. This article reviews Internet of Things, its different applications in several industries as well as its opportunities and challenges.

1.1 Introduction

Internet of Things (IoT) is the topic that we hear everywhere. IoT can be applied not only to many industries, but it comes directly to people's lives and affects them in a very positive fashion. IoT can not only revolutionize today products and services, but also it creates new business opportunities and markets.

We are living in a time period in which Internet is not only for connecting people, but now it is time to use internet for connecting objects together and to people (Nataliia and Elena 2015; Bandyopadhyay and Sen 2011). IoT is currently the most emerging technology which has passed its introduction phase, it is now in the center of attention (Gartner 2015) and will remain one of the most significant motives for evolving the

A. Razmi-Farooji (✉)
MyOpt Consulting Ltd, Metsätähdentie 7, 44200 Suolahti, Finland
e-mail: arian.razmi-farooji@myopt.fi

A. Razmi-Farooji
Industrial Engineering and Management, University of Oulu,
P.O. Box 4610, 90014 Oulu, Finland

P. Diez et al. (eds.), *Computational Methods and Models for Transport*,
Computational Methods in Applied Sciences 45,
DOI 10.1007/978-3-319-54490-8_1

information and communication technology in coming decades (Buckley 2006). In IoT vision, services and products can be monitored in a more visible and clear way through the whole value chain stages, including production, delivery and retail and they can communicate with other business sectors such as CRM, Vendor Managed Inventory and Business Analytics (Lee and Lee 2015).

It has been estimated that by 2025 IoT will affect the global economy with a value about 36 trillion dollars (Borgia 2014). That is why there is huge pressure from technological and even society side and there is a tough competition between companies on acquiring this technology. *Global Environment for Network Innovations* (GENI) and *Future INternet Design* (FIND) in the US, *Future Internet Research and Experimentation* (FIRE), run by European Union (EU), *Future Internet Forum* (FIF) in South Korea, the *New Generation Network Promotion Forum* (NWGN) in Japan (Ma 2011) and *Industrie 4.0* in Germany (Smart Service Welt Working Group 2015), are just some of research projects in international frameworks which focus on realizing IoT.

IoT paradigm is only possible to apply when its driver technologies are integrated together. Identification technologies such as RFID help a large group of objects get recognized. Thanks to networking and communication technologies, different and heterogeneous objects can connect and communicate to each other. Data analytics and signal processing enable analyzing large-scaled data which is gathered by sensing technologies, called as Wireless Sensor Networks (WSN) and stored in data storage.

Although implementing IoT is costly and needs huge investments, it makes companies to use their resources more efficiently (Lee and Lee 2015), opens up new opportunities—especially in Information and Communication Technology—leading to new products and services which can bring physical and virtual life together (Moretti et al. 2016). On the other hand, IoT affect individuals' lives through assisted living, digital health and home automation as well as industries via offering intelligent solutions for manufacturing, transportation, business and process managements (Atzori et al. 2010). The rest of this paper is organized as follows. Section 1.2 reviews briefly background and definition of Internet of Things. In Sect. 1.3 some of the most important application of IoT are discussed. Section 1.5 lists some challenges industries and academia have ahead in order to implement IoT paradigm. Finally Sect. 1.6 concludes the paper.

1.2 Definition and Background

The word "Internet of Things" was firstly coined by Kevin Ashton in a presentation he gave at Proctor and Gamble (P&G) in 1999 (Ashton 2009). Kevin Ashton, the brand manager at P&G, noticed that the shelves of a specific product became empty very soon and as a technical solution he started to research how to solve this problem by implementing RFID tags into P&G's supply chain system.

Although IoT paradigm has been introduced in early 2000s, the term "Internet of Things" has become official since the International Telecommunication Union

(ITU) published a report related to this subject in 2005 (Borgia 2014). Due to interdisciplinary nature of IoT and its huge and versatile applications, there are different definitions in the literature. While (Atzori et al. 2010) approaches IoT from three different viewpoints, namely Things oriented, focusing on various objects and ways to integrate them, Internet oriented which emphasize the important role of networks in such a system and Semantic oriented, considering information and knowledge management frameworks, (Borgia 2014) believes that definition of an IoT system can differ from purposes and interests that standard organizations, research centers and enterprises pursue and it evolves based on technologies and ideas over time.

Reviewing various definitions, available in the literature, an IoT-system can be defined as "An intelligent world-wide network, in which a huge number of heterogeneous objects, equipped with sensing and actuating capabilities are connected together and to people, share information and interoperate."

Products in such systems—known as Intelligent Products—can monitor their own condition, external environments and the way they are used, control their functions to the degree in which they provide a personalized user experience, optimize and improve their performance and react autonomously to the real—physical world (Porter and Heppelmann 2014). This "smartness" makes Intelligent Products to play an important role in business, information and societal environments, around which new services are defined. This forms the 6A Vision to IoT, stating that entities in an IoT-system are connected to anything and anyone which can be accessed anytime and from anywhere under any network and any service structure (Vermesan et al. 2011).

There are three main elements in an IoT-system: Hardware such as sensors and actuators which "things" are equipped with, Middleware, a cloud-based solution to store and analyze gathered data and Presentation tools which are used to show the data analytics results to the users (Gubbi et al. 2013). Combination of these three elements, realize an intelligent system through common understanding of the whole system, its users and the "things" which they are using, a comprehensive communication network which leads to processing and analyzing data by analytics tools (Gubbi et al. 2013). Figure 1.1 shows an IoT-system.

Having so many entities connected, integrated and communicated, a huge system, including several subsystems is created. This huge system is called System of Systems (SoS). In a SoS, the gathered data and information needs to be shared within all systems which besides having heterogeneous connected and integrated "things" makes the overall system, even more complicated. Airplanes and Space Shuttles are highly complicated products which can be categorized as Systems of Systems. A good example of IoT as a System of Systems is Smart Grid, Smart City, and Smart Home which are described in Sect. 1.3.1.

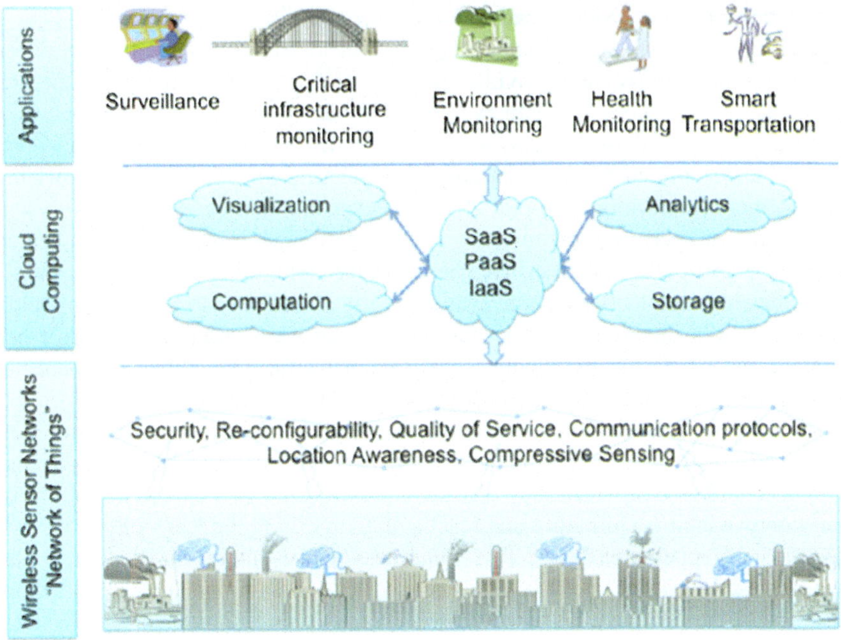

Fig. 1.1 Elements of an IoT System (Gubbi et al. 2013)

1.3 Applications

IoT has a huge range of applications in almost every industry and as IoT progresses, more applications are introduced. Following is a short glance at some potential applications around IoT, which both contain applications that are applied today in different industries and future directions.

1.3.1 Smart Environment

According to Mark Weiser, the father of ubiquitous computing, a smart environment is "the physical world that is richly and invisibly interwoven with sensors, actuators, displays, and computational elements, embedded seamlessly in the everyday objects of our lives, and connected through a continuous network" (Weiser et al. 1999). Although this definition is quite similar to the definition of IoT, nowadays smart environment refers to a set of IoT applications, such as Smart Grid, Smart City, Smart Home and Smart Factory (Atzori et al. 2010).

Smart Grid is an intelligent network of electrical systems which can store electricity, communicate bi-directionally to homes and make decisions, rather than traditional grids that are only capable of transferring and distributing electricity to central points (Tuballa and Abundo 2016) In Smart Grid, users not only consume electricity, but also they can even generate power, which as a matter of fact leads to a more efficient electricity transmission-distribution cycle and reduction in CO^2 emissions (Borgia 2014). Moreover, Smart Grids result to a more reliable and secure energy distribution, boost the share of generated electricity via renewable resources and shifting the consumption peak hours (Moretti et al. 2016).

In *Smart City* concept citizens, information and cities' physical infrastructural elements such as transportation, waste management and energy systems are all connected together, collecting information from the citizens and pass it to public and government in order to offer better services (Borgia 2014). Implementing the Smart City concept aims at solving problems such as traffic and pollution, resulting to better management of natural resources which as a matter of fact leads to sustainability and economic growth and a higher quality of life for the citizens (Lee et al. 2014).

Smart home, also known as *Smart Building* or *Home Automation* deals with having an interconnected system inside homes or buildings in which appliances are connected to each other and they interact with people. Firstly introduced in 1990s, Smart Building has a versatile set of applications such as automatically adaptable temperature inside buildings to residents' preferences or weather conditions outside, increasing buildings' safety through intelligent surveillance and alarming systems, managing lighting conditions based on different times of day and automatically switching on or off when people enter or leave the rooms (Atzori et al. 2010). Smart Building also plays an important role in Smart Grid concept by enabling communications between buildings and grids in a bi-directional fashion (Zhou et al. 2016). As previously mentioned, Smart Grid, Smart City and Smart home are a very good example of IoT as a system of systems.

Smart Factory includes a wide network of self-organizing smart objects, equipped with sensors actuators and processors which can all be uniquely identified within the factory by means of for instance RFID tags (Herrmann et al. 2014). Such Smart objects should be combined with big data analytics which provides higher degrees of efficiency and flexibility, enabling customized small lot manufacturing (Wang et al. 2016). Increasing Manufacturing flexibility through better utilization of resources and workers' skills (Borgia 2014), Smart Factory plays a significant role in smart production by providing high interconnection and integration of entities (Wang et al. 2016).

Other applications under Smart Environment which have been mentioned in the literature include *Smart Gym*, where tailored exercise can be uploaded to exercise machines as users enter the gym and identifies themselves by using RFID or NFC technology. In Smart Museum, as visitors walk through different section, sections can adapt themselves to the contents which visitors are watching like different historical periods or related information can be shown to them (Atzori et al. 2010).

1.3.2 Transportation and Logistics

Originated at first to solve a problem in P&G's supply chain, IoT greatly affects logistics, supply chain management and transportation. Thanks to RFID and NFC technologies, enterprises are able to monitor their entire supply chain at all stages including, procurement, transportation, warehousing, manufacturing, delivery process and even after-sale services on real-time bases (Atzori et al. 2010) where compared to the traditional tracking methods the state of products were static and could only be updated in different time periods (Jie et al. 2015).

This real-time monitoring of the entire stage helps supply chains to be more responsive where they can react to changes in customers' needs in shorter periods of time and run almost with zero safety stock (Bandyopadhyay and Sen 2011) as well as increasing the flexibility of suppliers, where they can react to changes in orders without any effects and other priorities and extending the lead-times, resulting to end customers' satisfactions (Jie et al. 2015).

IoT also enables retailers to track their products and items in real-time, helping them to monitor their inventory levels and prevent the out-of-stock, informing them when shelves inside retail stores become empty in order to replenish them and prevent their items to be stolen by shoplifters (Bandyopadhyay and Sen 2011). On the other hand this real-time monitoring by collecting data through wireless sensor networks or RFID tags, state of perishable products and cold chains like their temperature and humidity can be observed through the whole chain which checks whether the product has been spoiled or not and ensures the quality of deliverable products (Borgia 2014).

IoT can help enterprises with the right number of products to be produced and prevent under- or over-production. Gathering data from shelves helps retailers to understand the customers' shopping patterns and as a matter of fact helps them to produce as many as needed, leading to an optimized inventory system. Some globally well-known fast fashion retailers are benefiting from analyzing customers' requirements and shopping patterns, in order to produce what customers want as many as necessary and reduce the time-to-market just to two weeks (Christopher 2000).

IoT enables a better transportation planning and management by having a clear visibility over fleets which leads to more efficient vehicle routing and delivery process (Gubbi et al. 2013). On the other hand, this better management of transportation fleets results to reduction of carbon footprint supply chain fleets (Bandyopadhyay and Sen 2011). IoT provides a faster, more accurate and predictive monitoring capability over the cargo and containers meter-by-meter and second-by-second to make sure they reach to the right place on right time without being stolen (Macaulay et al. 2015).

Mobility is one of the most important aspects of urbanization. In IoT concept vehicles of all kind like buses, cars, trains as well as transportation infrastructures such as roads and rails will all assist drivers and passengers by providing them information related to roads safety, traffic patterns and weather conditions ahead of the road, navigate them through the best possible directions (Atzori et al. 2010). IoT facilitates the Connected Vehicles concept. In this concept, vehicles are connected to each other and their outside environments such as traffic lights and vehicles can communicate together and to external environments. Collision avoidance, lane change warning, green transportation because of smoother traffic and smart traffic scheduling are just some of benefits of connected vehicles concept (Lu et al. 2014).

1.3.3 Healthcare

IoT affects healthcare services greatly. The major applications of IoT regarding healthcare are related to automatic collection of patients' data, tracking people and objects, identification of personnel inside hospitals and remote monitoring of patients' health status.

Healthcare systems are large systems in which a lot of information related to doctors' and nurses' details as well as patients' records in different hospitals are being saved in different databases for detection, diagnosis and treatment. Since delivering such information at the right time to right place affects the quality of medical services, IoT facilitates fast access of information (Abinaya et al. 2015). With IoT, it is possible to collect data and transfer it to different stakeholders automatically, leading to shorter processing times (Atzori et al. 2010). The next application of IoT in medical services is tracking items and people inside a healthcare system. Equipping medical and surgical tools with smart labels can locate items wherever they are, prevents them to be stolen or to get lost (Borgia 2014) as well as preventing surgeons from leaving those tools unintentionally inside patients' body (Atzori et al. 2010).

Having medical equipment tagged with smart labels, hospitals can monitor their inventory on real-time basis and optimize it (Borgia 2014). Moreover, this approach can be also used for identifying the personnel inside a hospital and giving or not giving them access to different sections of hospitals (Atzori et al. 2010).

IoT enables real-time monitoring of patients' medical parameters such as blood pressure, body temperature, heart beat rate and sugar or cholesterol level in blood (Gubbi et al. 2013). Wireless Body Area Networks are miniaturized and intelligent wireless sensors which can be used around patients or implanted in their bodies in order to deliver a continuous flow of information regarding the patient's health status without any interference in his or her life (Yuce 2010). This technology helps doctors and other medical staff to monitor patients' health conditions remotely, as the patients are not in hospitals (Yuce 2010; Negra et al. 2016). This solution can be of special importance for those who are suffering from diabetics, cancer, coronary heart disease, stroke, Alzheimer's disease, cognitive impairments, seizure disorders and chronic obstructive pulmonary disease (Bandyopadhyay and Sen 2011).

Remote health monitoring can also be applied to monitor health conditions of elderlies. According to a survey in the United States, 30% of interviewees older than 65 years old mentioned that they prefer die, but not enter a nursing home (Mattimore et al. 1997). Although there are several use cases in which remote health monitoring has been applied in order to monitor cognitive and mental health as well as heart conditions in elderlies, the technology for having remote health monitoring in a smart environment such as smart homes is still not mature enough (Liu et al. 2016). In this sense, remote heath monitoring and smart homes are highly connected to each other, requiring more research from both industry and academia.

1.3.4 Fault Prediction and Predictive Maintenance

Maintenance costs can be between 15 to 50% of the whole annual production costs (Coetzee 2004) which shows such a crucial role maintenance plays. However, in today digital era, in which production systems and their components have become more inter-connected, the role of maintenance will be even more significant, because once a failure in one of production systems' components occur, the system as a whole will be affected and the value creation process will be hindered (Wang 2013; Denkena et al. 2012).

Maintenance in Digital Era is called E-maintenance. Coined in early 2000, E-maintenance is an interdisciplinary approach to asset management, which thanks to Information and Communication Technologies (ICTs) integrates and synchronizes the various maintenance and reliability functions, including monitoring, diagnosis, prognosis, decision and control process together and delivers asset information where and when it is needed (Muller et al. 2008).

The recent advances in ICT has transformed the traditional maintenance approaches in a way that they cannot answer today's requirements in the industry. Increasing rate of products and production systems complexity, lack of raw materials, the need for green and sustainable approaches, importance of knowledge creation (Aboelmaged 2015) as well as the need for a system which can consider the state of maintenance system in real-time and predict failures ahead of their occurrence are the major motives of having E-Maintenance (Mourtzis et al. 2016).

E-Maintenance is more than just a maintenance system. It is a strategic decision which changes the corporate's strategy towards maintenance from "fail and fix" to "predict and prevent" (Levrat et al. 2008). It integrates maintenance with other enterprises functions, makes product maintainable besides available and reliable which increases the products' lifecycles, enables cooperative and collaborative tasks as well as remote maintenance (Iung et al. 2009).

1.3.5 Other Applications

IoT applications are not limited to above-mentioned fields. There are more applications discussed in literature such as environmental monitoring and public safety which lets governmental authorities to offer better public security services to their civilians (Borgia 2014; Miorandi et al. 2012), Aerospace, Aviation, Automotive, Telecommunications and Pharmaceutical industry (Bandyopadhyay and Sen

2011), water network monitoring and other applications in Agricultural systems (Bandyopadhyay and Sen 2011; Gubbi et al. 2013) and Breeding (Bandyopadhyay and Sen 2011), as well as Insurance and Entertainment (Bandyopadhyay and Sen 2011).

1.4 Opportunities

In addition to above mentioned applications, IoT also opens up new opportunities, both for individuals and economy as a whole. In order to benefit from opportunities that IoT offers, firstly it is important to develop a comprehensive understanding of business and legal issues. The versatile nature of IoT which combines and integrates different technologies forms new business models, in which new methods, tools, software and services are created. New business models, offered by current players or new comers focus on creating new services based on data and information, collected from sensors and actuators or mobile devices such as smart phones (Peña-López et al. 2005).

Like internet which offered new opportunities when it was commercialized, IoT is a competitive advantage which transforms the competition in ways that just traditional market leaders and competitors are not competing anymore, but new and innovative ideas in form of start-ups also enter the competition (Friess 2013). This new business generations affect the economy as in larger scale. For instance it has been predicted that in 2025, the US GDP increases by 2–5% which leads to creation of more jobs and increasing the quality of individuals' lives (Borgia 2014).

1.5 Challenges

Despite of variety of applications and new opportunities that IoT can offer, its implementation is challenging. Following is a list of mostly discussed challenges in the literature which are still controversial and both academia and industry are putting effort in order to solve them.

1.5.1 Security

A research by Hewlett Packard shows that 70% of devices used in an IoT systems are vulnerable to attacks (Meissler 2014). IoT is a huge system of interconnected systems with large number of heterogeneous and smart objects which are connected together through wireless networks such as internet. Therefore, so large-scaled system such as IoT intensify the security threats. Security requirements in context of IoT can be referred to as authentication, confidentiality and access control (Sicari et al. 2015).

Moreover, most of components in an IoT system are wireless, left unattended, and are low capable in sense of energy and computing resources which make them really assailable (Ashton 2009).

In scope of security, the main challenges are related to authentication, confidentiality and access control (Sicari et al. 2015). Data authentication is a major principle which deals with authenticating the retrieved address and object information (Borgia 2014). However, authentication is a challenging task, since it needs specific authenticating infrastructure like servers which are able to communicate with other nodes while some components such as passive RFID tags are not adequately able to communicate with servers (Atzori et al. 2010).

In order to achieve confidentiality, encryption and decryption technologies must be implemented. However the traditional encryption techniques cannot be scaled up to IoT systems, because devices in an IoT environment are source-constrained (Mao et al. 2016) Therefore there are more efficient encryption techniques necessary which are quick in sense of computation processes and consume less energy (Bandyopadhyay and Sen 2011).

1.5.2 Standardizations

IoT is a huge ecosystem, in which not only a variety of industries but also citizens and individuals are involved. Therefore, standardization is absolutely necessary to guarantee the interoperability of different IoT components, applications, industries and services. The main focus of standardization is to focus to remove technical obstacles and challenges such as data format, data interfaces, service platforms and architecture in order to ensure that different smart objects with versatile natures can communicate and interoperate (Borgia 2014).

There are quite many initiatives are working on unifying the IoT-standards. International Standard bodies (ISO and ITU), European Commission and European Standard Organizations (e.g. ETSI, CEN, CENELEC) as well as different departments in Auto-ID lab and EPC global (Atzori et al. 2010) are working on standardizing communication protocols, RFID, sensors and NFC technologies (Borgia 2014).

1.5.3 Addressing

IoT is a system of large number of nodes, in which information from every available node should be retrievable uniquely. Therefore, naming, identification and uniquely addressing different components is one of the major requirements in such a system. Addressing is used in order to identify different and heterogeneous smart objects like computers, laptops, wearables, smart phones and so on in a way everyone regardless of his or her position can retrieve the content generated by such devices (Borgia 2014; Atzori et al. 2010).

Since Internet Protocol (IP) is a must when it comes to any internet connection, one existing solution for addressing and networking in IoT system is assigning a dedicated IP to every object in the system (Palattella et al. 2014). Introduction of Internet Protocol version 6 (IPv6) has solved this problem to some extents, since IPv6 provides an unlimited number of unique addresses as 2^{128} which has outperformed the previous version IPv4 with 4 billion unique addresses (Palattella et al. 2014). However, there is still a lack of standard architecture for Naming, Addressing and Profile Server (NAPS) as a middleware which can be interoperable with different platforms (Liu et al. 2014).

1.5.4 Privacy

While confidentially is important for businesses, personal and private information of users on how and they are acting in the IoT environment must be protected. As it was mentioned earlier, IoT has a broad range of applications such as healthcare, smart home, connected cars and smart shopping where securing users' private information is of absolute importance.

Privacy has been defined as hiding the personal information and the ability of controlling how that information is being used (Weber 2010). Taking this fact into account that huge pile of information is collected and mined from individuals, people should know that who is collecting what data and for what purpose (Ashton 2009).

1.5.5 Other Challenges

Other challenges stated in the literature are Network Foundation and Managing heterogeneity of objects (Bandyopadhyay and Sen 2011), data and chaos Management (Lee and Lee 2015) IoT architecture, energy efficient sensing, quality of service, Geographic Information System (GIS) based visualization (Gubbi et al. 2013) as well as distributed systems technology and distributed intelligence (Miorandi et al. 2012).

1.6 Conclusion

Internet of Things has revolutionized industries and academia since its introduction. IoT has a variety of applications in almost in any fields which help enterprises use their resources more efficiently, reduce costs and increase their productivity. On the other hand, IoT has opened up new business opportunities, generated new business models and transformed the ordinary and traditional product-based industry to more service-driven world. Introducing new opportunities changed the business environments in a way that traditional companies are not the only players in the market and new ideas are generated in form of start-ups which create more jobs and affect the economy as a whole. However, implementing such a huge system of systems is challenging. Privacy, security, addressing and naming issues as well as standardizations are among the biggest challenges ahead for fully implementation of IoT.

References

Abinaya, Kumar V, Swathika (2015) Ontology based public healthcare system in internet of things (IoT). Procedia Computer Science 50:99–102

Aboelmaged MGS (2015) E-maintenance research: a multifaceted perspective. J Manuf Technol Manag 26(5):606–631

Ashton K (2009) That 'internet of things', thing. RFiD J 22(7):97–114

Atzori L, Iera A, Morabito G (2010) The internet of things: a survey. Comput Netw 54(15):2787–2805

Bandyopadhyay D, Sen J (2011) Internet of things: applications and challenges in technology and standardization. Wirel Pers Commun 58(1):49–69

Borgia E (2014) The internet of things vision: key features, applications and open issues. Comput Commun 54:1–31

Buckley J (2006) From RFID to the internet of things: pervasive networked systems. In: Final report on the conference organized by DG information society and media, networks and communication technologies, brussels

Christopher M (2000) The agile supply chain: competing in volatile markets. Ind Mark Manag 29(1):37–44

Coetzee JL (2004) Maintenance. Trafford, Bloomington

Denkena B, Bluemel P, Kroening S, Roebbing J (2012) Condition based maintenance planning of highly productive machine tools. Prod Eng 6(3):277–285

Friess P (2013) Internet of things: converging technologies for smart environments and integrated ecosystems. River Publishers, Denmark

Gartner I (2015) Gartners 2015 hype cycle for emerging technologies identifies the computing innovations that organizations should monitor

Gubbi J, Buyya R, Marusic S, Palaniswami M (2013) Internet of things (IoT): a vision, architectural elements, and future directions. Futur Gener Comput Syst 29(7):1645–1660

Herrmann C, Schmidt C, Kurle D, Blume S, Thiede S (2014) Sustainability in manufacturing and factories of the future. Int J Precis Eng Manuf Green Technol 1(4):283–292

Iung B, Levrat E, Marquez AC, Erbe H (2009) Conceptual framework for e-Maintenance: Illustration by e-maintenance technologies and platforms. Annu Rev Control 33(2):220–229

Jie YU, Subramanian N, Ning K, Edwards D (2015) Product delivery service provider selection and customer satisfaction in the era of internet of things: a Chinese e-retailers perspective. Int J Prod Econ 159:104–116

Lee I, Lee K (2015) The internet of things (IoT): applications, investments, and challenges for enterprises. Bus Horizons 58(4):431–440

Lee JH, Hancock MG, Hu M-C (2014) Towards an effective framework for building smart cities: lessons from Seoul and San Francisco. Technol Forecast Soc Chang 89:80–99

Levrat E, Iung B, Marquez Crespo A (2008) E-maintenance: review and conceptual framework. Prod Plan Control 19(4):408–429

Liu CH, Yang B, Liu T (2014) Efficient naming, addressing and profile services in internet-of-things sensory environments. Ad Hoc Netw 18:85–101
Liu L, Stroulia E, Nikolaidis I, Miguel-Cruz A, Rincon AR (2016) Smart homes and home health monitoring technologies for older adults: a systematic review. Int J Med Inf 91:44–59
Lu N, Cheng N, Zhang N, Shen X, Mark JW (2014) Connected vehicles: solutions and challenges. IEEE Int Things J 1(4):289–299
Ma H-D (2011) Internet of things: objectives and scientific challenges. J Comput Sci Technol 26(6):919–924
Macaulay J, Buckalew L, Chung G (2015) Internet of things in logistics. A collaborative report by DHL and Cisco on implications and use cases for the logistics industry, Pub. DHL Customer Solutions & Innovation, Troisdorf
Mao Y, Li J, Chen M-R, Liu J, Xie C, Zhan Y (2016) Fully secure fuzzy identity-based encryption for secure IoT communications. Comput Stand Int 44:117–121
Mattimore TJ, Wenger NS, Desbiens NA, Teno JM, Hamel MB, Liu H, Califf R, Connors AF, Lynn J, Oye RK (1997) Surrogate and physician understanding of patients' preferences for living permanently in a nursing home. J Am Geriatr Soc 45(7):818–824
Meissler D (2014) HP study reveals 70 percent of internet of things devices vulnerable to attack. Retrieved 30 June 2015
Miorandi D, Sicari S, De Pellegrini F, Chlamtac I (2012) Internet of things: vision, applications and research challenges. Ad Hoc Netw 10(7):1497–1516
Moretti M, Djomo SN, Azadi H, May K, De Vos K, Van Passel S, Witters N (2016) A systematic review of environmental and economic impacts of smart grids. Renew Sustain Energ Rev
Mourtzis D, Vlachou E, Milas N, Xanthopoulos N (2016) A cloud-based approach for maintenance of machine tools and equipment based on shop-floor monitoring. Proced CIRP 41:655–660
Muller A, Marquez AC, Iung B (2008) On the concept of e-maintenance: review and current research. Reliab Eng Syst Saf 93(8):1165–1187
Nataliia L, Elena F (2015) Internet of things as a symbolic resource of power. Proced Soc Behav Sci 166:521–525
Negra R, Jemili I, Belghith A (2016) Wireless body area networks: applications and technologies. Proced Comput Sci 83:1274–1281
Palattella M, Latif L, Ziegler S, Kastner W, Jung M, Kofler M (2014) IoT-IPv6 integration handbook for SMEs
Peña-López I et al (2005) ITU Internet report 2005: the internet of things
Porter ME, Heppelmann JE (2014) How smart, connected products are transforming competition. Harv Bus Rev 92(11):64–88
Sicari S, Rizzardi A, Grieco LA, Coen-Porisini A (2015) Security, privacy and trust in internet of things: the road ahead. Comput Netw 76:146–164
Smart Service Welt Working Group/acatech (eds) (2015) Smart service welt—recommendations for the strategic initiative web-based services for businesses. Final Report. Berlin, March 2015
Tuballa ML, Abundo ML (2016) A review of the development of smart grid technologies. Renew Sustain Energ Rev 59:710–725
Vermesan O, Friess P, Guillemin P, Gusmeroli S, Sundmaeker H, Bassi A, Jubert IS, Mazura M, Harrison M, Eisenhauer M et al (2011) Internet of things strategic research roadmap. Int Things Glob Technol Soc Trends 1:9–52
Wang L (2013) Machine availability monitoring and machining process planning towards cloud manufacturing. CIRP J Manuf Sci Technol 6(4):263–273
Wang S, Wan J, Zhang D, Li D, Zhang C (2016) Towards smart factory for industry 4.0: a self-organized multi-agent system with big data based feedback and coordination. Comput Netw 101:158–168
Weber RH (2010) Internet of things-new security and privacy challenges. Comput Law Secur Rev 26(1):23–30
Weiser M, Gold R, Brown JS (1999) The origins of ubiquitous computing research at PARC in the late 1980s. IBM Syst J 38(4):693

Yuce MR (2010) Implementation of wireless body area networks for healthcare systems. Sens Actuators A Phys 162(1):116–129

Zhou B, Li W, Chan KW, Cao Y, Kuang Yonghong, Liu Xi, Wang Xiong (2016) Smart home energy management systems: concept, configurations, and scheduling strategies. Renew Sustain Energ Rev 61:30–40

Chapter 2
A Computational Modeler's Tour of the Port of Houston

Niels Aalund and William Fitzgibbon

Abstract When one thinks of Houston a variety of different economic activities might come to mind: energy, space, and medicine. Typically one does not think first of transportation. However, the transportation enterprise in the Houston area is massive. Houston is one of the few places in the world where the various modes of transportation: rail, highway, maritime, inland waterway, pipeline and air converge. Approximately 90% of the volume and 70% of the value of goods transported worldwide takes place at sea. These general percentages also apply for the Houston region. Our focus will be on the maritime sector in particular the Port of Houston and the Houston Ship Channel.

2.1 Introduction

The Port of Houston is a 25 mile complex comprised of public terminals and more that 150 private industrial sites along the 52 mile Houston Ship Channel. It stretches from close to downtown Houston to Galveston Bay. In terms of international waterborne tonnage the Port of Houston ranks first in the United States (US) and second in the US in terms of total cargo handled. The Port has the second largest petrochemical complex in the US. It refines over 35% of the nation's gasoline and 80% of it aviation fuel. The annual economic impact of the Port of Houston is estimated to be over 178.5 billion dollars (Port of Houston Authority 2016). Given the enormity of the economic activity: environmental, safety, and security concerns are all of paramount importance. Even a short term interruption of activity results in a loss of millions if not billions of dollars. Any protracted disruption of traffic on the Houston Ship

N. Aalund (✉)
Western Gulf Maritime Association, 1717 Turning Basin Dr #200, Houston, TX 77029, USA
e-mail: naalund@wgma.org

W. Fitzgibbon
Departments of Engineering Technology and Mathematics, University of Houston,
Houston, TX 77204, USA
e-mail: wfitzgib@central.uh.edu

P. Diez et al. (eds.), *Computational Methods and Models for Transport*,
Computational Methods in Applied Sciences 45,
DOI 10.1007/978-3-319-54490-8_2

Channel and its more than 150 port facilities would have a catastrophic impact on the U.S. economy. The Houston Area is subject to extended annual fog closings of the Houston Ship Channel. These fog closures catch the attention of national and international media sources. Computational science and mathematical modeling may not normally be associated with maritime activity. In what follows, we shall use the Port of Houston/Houston Ship Channel as a lens through which computational and mathematical scientists can view their operations, environmental concerns, and security issues. Through this one can determine that modeling, computation, simulation and analysis can and do play an important role in the maritime enterprise.

2.2 Traffic Operations

In many ways the management of traffic in and out of the Port of Houston via the Houston Ship Channel resembles the scenario of air traffic control at a major airport. Over 11,000 vessels (blue water) visit the Port of Houston. This breaks down into a daily average of over 300 vessel arrivals, a figure exceeding the daily arrival of passenger planes at Stockholm Arlanda Airport. The Houston Ship Channel is also one of the nation's busiest barge transportation lanes (brown water). It intersects with the Intracoastal Waterways which runs for over 3,000 miles from Boston to the Rio Grande connecting Houston with the Eastern Seaboard of the United States. In additional it connects the Midwest and South Central US via the Mississippi/Missouri, Ohio/Allegany and Tennessee/Tombigbee Rivers. Added to this blue water and brown water traffic there is a large fishing and shrimping fleet, massive recreational craft activity and cruise ship operations in and out of the both the Port of Galveston and the Port of Houston.

Vessels operating on the high seas and on the coastal and inland waterways are monitored by Automatic Identification Systems (AIS). AIS technology gathers and transmits data such as vessel identification as well as track and monitor position, course, and speed. The transmitted information is displayed on radar like screens on board other vessels and on land based operations center such as the Port of Houston. Screens at land based operations centers such as the Port of Houston provide real time representation of regional maritime activity and in many ways resemble what one sees at an air traffic control center. AIS information is primarily for traffic monitoring and collision avoidance and to provide information about a vessel and its cargo to local authorities who oversee waterborne trade (Board 2003). The U.S. Coast Guard Vessel Traffic Services (VTS) depends heavily on AIS systems in their crucial role monitoring and controlling vessel traffic.

2.3 Scheduling and Logistics

Industry is driven by profits and profit margins. Consequently, it faces an ongoing challenge of optimizing supply chain networks. Supply chains have become global, making maritime shipping the primary link within many supply chains. It is therefore understandable that a critical component of supply chain optimization is a proper scheduling methodology. It has been estimated that a 7% annual reduction in logistics costs in refining translates to 23% increase in profitability (Burns 2014). Despite the fact that the dominant proportion of the transportation of goods worldwide is maritime, maritime scheduling has been studied much less extensively than rail, truck or plane transport (Burns 2014). Although these modes of transportation have similarities in terms of scheduling, ship fleet planning problems are differentiated from those of other modes of transportation (Al-Hamad et al. 2012): ships operate continually around the clock without buffers of planned down time that can absorb delays; they have higher uncertainty in their operations due to their higher dependence on weather conditions, technology and operating over long distances with long time horizons. Vessels frequently cross multiple national and international jurisdictions. Typically they carry mostly liquid and dry bulk cargo, and often nonmixable products in separate compartments (Christiansen et al. 2007).

Inventory Routing Problems (IRP) which integrate inventory management with maritime vessel routing and delivery have been a subject of active study for the past thirty years. Prior to this period despite an awareness and an interest in distribution and inventory problems little had been done to integrate the two. This can be attributed to previous limitations on computational capacity and the association of non-availability of advanced algorithms, capable of handling large and complex combinatorial problems (Coelho et al. 2013). IVP's can be static or dynamic depending upon whether or not all information is known in advance or is revealed over time. Depending upon the certainty of the information the problems can be deterministic or stochastic.

The Port of Houston accommodates a variety of different types of cargo: bulk cargos which haul oil, chemicals and grains; intermodal container cargos, and break bulk cargo that must be loaded individually. Common examples of break bulk cargos include: oil field equipment, automobiles, steel, wind turbine propellers, bailed goods, heavy equipment, and drums. Each of these cargos present a different set of unique scheduling and logistics problems. However, each must be coordinated with dockside scheduling at the Port of Houston.

Refinery and petrochemical plant operations involve: marine terminals, process terminals and pipelines. Inventory levels depend upon process scheduling, demand at the depots and availability of pipelines. The complexity of the problem becomes apparent with the realization that realistic optimization models need not only to account for process scheduling at the refineries, demand at the depots and pipeline availability but they also have to be able to account for variability in transit time and uncertainty of demand over a time horizon greater than 40 days (Persson and Göthe-Lundgren 2005).

Issues unique to port container terminal operations are receiving more attention due to increasing importance of marine transportation systems. The most important objective for a port container terminal is to increase its throughput or, in particular, to decrease the turnaround times of ships (Al-Hamad et al. 2012). This entails optimizing turnaround time and effective of allocating and scheduling of key resources, such as berths, yards, quay cranes, yard cranes and trucks (Ng and Mak 2005). Break bulk cargo handling is more labor intensive than either container handling or the handling of bulk cargos such as crude oil or refined products. Although break bulk cargos can at times be loaded onto vessels directly from rail cars, trucks, or discharged from vessels directly to the chosen mode of transportation. They are commonly delivered to dockside warehouses prior to the arrival of a vessel and taken from the warehouses to the quays for loading. Similarly discharged cargos are usually unloaded onto quays and taken to warehouses or storage areas where they await transfer to trucks or rail. Rail and truck scheduling and routing issues have been studied extensively (Hwang 2005). However, warehouse scheduling has remained a challenge. It is a complex combinatorial optimization problem with various constraints whose computational complexity increases exponentially with scale (Yang et al. 2013).

2.4 Security and Safety Issues

Goods from countries all over the world come into the Port of Houston making it (along with other major seaports worldwide) a tempting target for the launch of a major terrorist attack with devastating consequences.

2.4.1 Physical Security

Ports contain a number of specific facilities that could be targeted by terrorists. In addition to vessels and infrastructure, terrorists may seek to attack maritime communities using ships or containers on ships as delivery vehicles for weapons of mass destruction or by exploiting chemicals, explosives in cargo ships, onshore industrial operation, and storage tanks in populated port area (Parfomak and Frittelli 2007). A well-organized and implemented attack on the Port of Houston could inflict a high number of causalities and cause catastrophic economic damage of to both the nation and world economies. Maritime attacks have occurred in a variety of ports worldwide and terrorists have a variety of options at their disposal. Terrorism threat in the maritime domain may come in various forms. Weapons or explosives may be concealed in containers, ships may be used as weapons to destroy critical infrastructure, or terrorists may illegally cross the borders to launch attacks in the homeland. Much emphasis has been has been placed on the security of containerized cargo however it is important to recognize other vulnerabilities as well (Bakir 2007).

The Houston Ship Channel is spanned by three major bridges. Each of these structures is subject to damage resulting from collisions with ships or barges, storm, or even terrorist activity (for example the Brooklyn Bridge and possibly the Golden Gate Bridge, have been unsuccessfully targeted). Such occurrences would result in and in some cases have resulted in a shutdown of the channel or portions thereof.

2.4.2 *Plant Explosions*

There are about two major explosions around the world every year in the refinery and petrochemical industry (Lewis 2006). Petrochemical plant and refineries are inherently hazardous due to the materials that they handle and their manner of operations. There is always a risk of explosions, fire, and the release of toxic substances (Khan 1999). Accidental scenarios in refineries, petrochemical and other installations handling flammable chemicals in bulk include pool fire, jet fire, flash fire, boiling liquid expanding vapor explosions and vapor cloud explosions. Individuals and communities can be injured by the effects of the fire, the explosion, and by the release of chemicals (Birwas 2000). There is a mandate for the enhancement of safety in the design or the designs of renovations of these facilities but design must be accomplished within the constraints of an appropriate level of cost. Here simulations can play an important role. They serve as a tool for optimal choice of design while addressing safety concerns within the constraint of an appropriate level of cost. Numerical Simulations based on Computational Fluid Mechanics (CFD) can be used in a wide of facilities including refineries, petrochemical plants, gas processing plants and LNG plants to predict gas dispersions, blast pressures both near and far field and structural responses (Takahashi and Watanabe 2010). A wide variety of CFD based models for gas explosions exist each with its own strengths and weaknesses. Improvements in both the physics and the numerics are required for greater confidences in the predictions of the models (Lea and Ledin 2002).

2.4.3 *Cyber Security*

In addition to conventional physical attacks there is a rapidly growing threat of cyber-attacks (Obama 2013). US ports are very vulnerable to cyber-attacks. US ports rely as much upon networked computers and control systems as they do upon shore-side labor. However, it is both surprising and disconcerting that little attention has been paid to the security of networked systems that undergird port operations and no cybersecurity standards for US ports have been put in place (Kramek 2013). Networked systems are involved at all stages of maritime transport: they track maritime cargo from the time a container is filled overseas until it reaches its final destination, networked control systems are employed involved in cargo loading and discharging. Cranes and other systems use optical recognition along other technologies to locate, scan, and manage aspects of port terminal operations. Scanners and radio frequency identification devices (RFID) track cargoes and as well as trucks and rail car entering and exiting ports. The U.S. Coast Guard has identified cyber-security vulnerability as a major concern and have initiated proactive dialogue with the maritime industry.

The Port of Houston has a sophisticated fiber optic network of security cameras, radars, sonar sensors, and other systems which continually monitor the progress of a vessel up and down the Houston Ship Channel. Networked systems optically scan cargo and manage its movement throughout as well as into and out of Port of Houston facilities. Cranes, fuel farms, HVAC systems, and access control systems are monitored. However, the emphasis is upon physical not cyber security (Kramek 2013). The situation is further complicated by the fact that Port of Houston authorities lack cybersecurity knowledge of most of the privately owned port facilities of oil, gas and chemical tanker ships. The oil and gas and petrochemical industries utilize networked systems govern that all aspects of their operations. What, if any, cybersecurity measures these privately owned facilities have in place is mostly known only to the entities that own and operate these facilities. They work independently in an autonomous environment. SCADA systems abound in production, refining and petrochemical plant operations creating a valid concern that a compromise of SCADA system could produce a major spill or explosion (Bronk 2014). The cyber threat comes from a variety of sources: terrorists (international or domestic); activists, pressure groups, single-issue zealots; disgruntled employees or contractors; and criminals (e.g., white collar, cyber hacker, organized opportunists) (American Petroleum Institute 2003).

The U.S. Coast guard requires all vessels to maintain a vessel security plan. Likewise, the U.S. Coast Guard also requires all terminals to maintain a facility security plan (FSP). Information security threats exist on board the vessels as well dockside. Given its primary importance and prevalence in maritime traffic safety, it is essential that AIS information be secure. In this regard (Balduzzi et al. 2014) recently conducted a comprehensive survey of AIS from both the soft and hardware perspectives. They were able to identify several specific threats affecting AIS implementation and protocol specifications. These include disabling AIS communications; tampering with existing AIS data; triggering alerts to lure ships into navigating to hostile sea space; or spoofing collisions to possibly bring a ship off course.

2.4.4 Work Place Safety

Work accidents and safety programs to prevent them are expensive. Indeed the National Safety Council estimated (2002) that economic loss from workplace accidents was almost \$150 billion dollars a year (do Carmo et al. 2010). Ports are a dangerous places to work. Dock workers face many hazards during the loading and unloading of cargoes as well as from movement and operations of vehicles. Moreover, most accidents in ports are serious or fatal. Typical workplace transport hazards are: loading and unloading, reversing, driving on dockside or in container storage areas, falls, coupling and uncoupling on dockside and on the ship. Container operations present the risk of accidents from specialized lifting equipment such gantry cranes, transtainers, slewing cranes and special fork lift trucks (AMEPA 2008).

Safety programs can be described as a dynamic set of interventions implemented at the work site for the purpose of reducing the likelihood of accidents and/or reducing the severity of such incidents should they occur (do Carmo et al. 2010). Safety interventions may and do include training, meetings, protective gear, equipment modifications, inspections, changes in procedure, and safety monitoring. Because they can be expensive, time consuming, and disruptive it is important to have assurance a priority of their effectiveness. Resources are always limited and consequently there is a need to develop optimizing strategies to decrease accidents at minimal cost. This is both a human resource development problem and a statistical forecasting problem. Statistically one can approach the problem with the systematic collection of data and the application of statistical forecasting methodology to identify a time lagged mathematical relationship between the implementation of interventions and the number of incidents (do Carmo et al. 2010).

2.5 Environmental Issues and Threats

The Port of Houston has many environmental challenges and threats. Some are anthropogenic, some are natural and some are combination of the two. Resolution of these issues and mitigation of the threats will present computational challenges and interesting opportunities.

2.5.1 Hurricanes

Hurricanes are an annual threat to the Port of Houston and the 150 other private facilities located along the Houston Ship Channel. Hurricanes are defined as rapidly rotating storm systems characterized by a low-pressure centers, strong winds, high waves, storm surges and a spiral arrangement of thunderstorms that produce heavy rain. Although the wind and rainfall associated with hurricanes is fearsome and capable generating significant damage, the Galveston Bay region which includes the Houston Ship Channel is extremely vulnerable to storm surge flooding. In 2008, the land fall of Hurricane Ike made landfall slightly east of the Houston Ship Channel and produced a storm surge that flooded inland areas stretching from Galveston Bay to well over 100 miles eastward. Hurricanes are rated in terms of increasing intensity on a scale from 1 to 5. Ike was "only" a Category 2 storm yet it caused over $25 billion dollars in damages. Research indicates that had Hurricane Ike made landfall 30 miles southwest of its original landfall location, it would have generated 18 ft of surge in the Houston Ship Channel and 15 ft inundating the populated areas, industries and facilities areas on the west side of Galveston Bay. Thousands of people who chose not to evacuate could have died. The environmental impact to Galveston Bay and economic devastation to the Houston-Galveston region and US would have been enormous (Blackburn et al. 2014). It is highly probable that a Category 3 or 4 winds will come ashore on the western side of Galveston Bay. If this occurs one can expect a 25 to 30 foot storm surge. If the Houston-Galveston Area has failed to protect critical infrastructure against a 25-foot to 30-foot surge, a surge of this magnitude could literally destroy the economy of the Houston region, if not the United States. In addition to the economic disaster, such a surge would likely inflict massive environmental damage if the hazardous materials and oil presently stored adjacent to the ship channel were to spill into adjacent neighborhoods as well as Galveston Bay (— 2010).

Hurricanes are inevitable and nothing can be done to lessen the amount of rainfall or dampen the force of the winds. However, there is the potential to mitigate the effects of storm surges by the construction a system including: dams, floodgates, and levees. Major storm surge projects have been constructed in Holland, the UK and Russia. Constructing such a system to protect the Houston Ship Channel would be a massive engineering project costing billions of dollars and it cannot proceed in a hit or miss fashion. The design of the system will require the development of robust, realistic computational models validated by historical data of storm surge, wave action and costal not only for the analysis of future risk and real time forecasting but perhaps most importantly for the optimal design of infrastructure to mitigate the impact of future events. In the aftermath of Hurricane Ike and its predecessor, there have been significant advances in numerical modeling of storm surge, wave, and coastal flooding. We currently have at our disposal a variety of models incorporating different physics and numerics. Most of these models are two or three dimensional incorporating the physics, complex bathymetry and topography of the domain as well as tide and wave action however, there is significant variance in their computational speed, domain size and resolution. Uncertainties of storm surge and coastal flooding calculations come from a variety of sources, including model physics, model domain and grid, model dimensionality, parameterization of model processes (e.g., surge-wave interaction, interaction with topographic features), and input data (model forcing such as wind, precipitation, tides), open boundary condition and bathymetry and topography data (Blackburn et al. 2014)

2.5.2 *Air Quality Issues*

Air pollution is a problem in urban areas worldwide and the Greater Houston-Galveston, Texas region, is no exception. The emissions from ships, diesel trucks, locomotives and other equipment coupled with emissions from the industries along the Houston Ship Channel have been identified as the major contributors to Houston's problem. Air pollution levels in Houston have been monitored for the last four decades. The Texas Commission on Environment Quality reports that air quality in Houston is monitored more closely and analyzed with more intensity than perhaps anywhere in the country (Bethel et al. 2006) with more monitors tracking over 130 chemical pollutants. However, the Houston air quality monitoring network is primarily designed to track levels the six pollutants (ozone, particulate matter, carbon monoxide, sulfur dioxide, lead, and nitrogen dioxide) for which and The United States Environment Protection has established health based standards.

Houston's most pressing and major problem is ozone. Houston meets the EPA standards for 5 of the 6 pollutants on the list, but does not meet the standards for ozone. Regretfully, the local situation regarding high levels of troposphere ozone concentrations in the Greater Houston Galveston has become recognized as being among the nation's worst (Bethel et al. 2006; Rivera et al. 2014). Consequently, Houston finds itself mandated by the Federal Clean Air Act of 1970 (amended 1990) to meet EPA standards. The Clean Air Act further stipulates that computer modeling be used to plan air pollution reduction. These computational models need to be validated by historical data obtained from the air quality monitoring system. Ozone is not released directly into the atmosphere but is created by the photochemical reaction of substances such as volatile hydrocarbons and the oxides of nitrogen sulfur that are released anthropogenic, from industrial complexes, mobile sources and biogenic from natural sources. The rate of product is highly effected by factors such as the amount of sunlight, temperature and the humidity. Area concentrations are dependent upon wind patterns as well. The computer models rely upon simulations based upon the computation of solutions to complex systems of partial differential equations to simulate emissions, air movement, chemical reactions, and resultant ozone concentrations. The basic mathematical equations describing the atmospheric production and distribution of ozone and other pollutants are weakly coupled system of reaction advection diffusion equations. This system incorporates a variety of processes such as chemical reactions, temperature, condensation processes as well as transport, thermal, and turbulent diffusion. Emissions become part of the boundary data. Even simple prototypes can involve over twenty equations (Stockwell et al. 2011) and a realistic model would couple hundreds of equations (Fitzgibbon and Langlais 2002). Once confidence of the validity of the models is obtained, the task becomes one of modeling various methods of abatement of emissions sources on the boundary in such a way as to bring ozone concentrations down to a level of attainment of the EPA standard.

2.5.3 Biological Invasions

Without entering into the esoteric controversy of the precise scientific definition we will use an online definition of biological invasion as: the process by which species (or genetically distinct populations), with no historical record in an area, breach biogeographic barriers and extend their range. Non-indigenous organisms which have been introduced into ecosystems are the second greatest cause, after habitat destruction, of species endangerment and extinction worldwide. In the United States, nonindigenous species have been estimated to do more than $130 billion annually to agriculture, forests, rangelands, and fisheries (Schmitz and Simberloff 2001). Maritime transport is also widely recognized as a primary if not the dominant invasion pathway by which non indigenous animal and plant species are introduced to ecosystems (Miller et al. 2011). Indeed the role of shipping in the biological invasion of North America has its antecedent in the early 17th Century with the inundation of New England and the Mid-Atlantic by a wave of rats, insects, mice, and aggressive weeds from Europe. In the 1930's Fire Ants from South America were introduced to North America via the Port of Mobile (Caldera et al. 2008). Today, several thousand nonindigenous species are established and new potentially invasive species arrive every year. For example, in Texas, an exotic snail carries parasites that are spreading and infecting native fish populations. In the Gulf of Mexico, a rapidly growing Australian spotted jellyfish population is threatening commercially important species such as shrimp, menhaden, anchovies, and crabs (Schmitz and Simberloff 2001).

Biological invasions have accelerated in recent decades putatively as a result of the dramatic growth of global trade and transport. Invasive species can be transported by a variety of different mechanisms including: cargo, hull fouling, and the loading and discharge of ballast water. Ballast water taken in or discharged by large ships to maintain stability is assumed to represent the world's largest invasion vector (Caldera et al. 2008). Ballast water from cargo ships has been implicated in transporting a South American strain of cholera to the Gulf of Mexico, leading to fish and shellfish contamination. Ballast water discharge is also responsible for the introduction of "red tide" algae to the waters of several countries, contaminating shellfish and threatening human health.

It is possible to employ sophisticated genetic and statistical methodology to identify the geographic source of invading non-indigenous species ex post facto (Seebens et al. 2013). However, we do need to have eradication efforts to control invading species and we need to have prophylactic measures in place that limit their ability to cross geo-barriers. We also need to develop robust, comprehensive predictive models of biological invasions that take into account ship routing, port environments and biogeography (Caldera et al. 2008).

2.5.4 *Infectious Disease*

The spread of infectious disease is closely related to the concept of bioinvasions. Long distance transport of pathogens is a major factor in the global spread of infectious disease. Initially, new infectious diseases could spread only as fast and far as people could walk. Then it was spread as fast and far as horses could gallop and ships could sail. With the advent of truly global travel and the globalization of world trade we have seen more new diseases than ever before become potential pandemics (WHO 2004).The World Health Organizations reports that one hundred outbreaks of infectious diseases were reported to be associated with ships between 1970 and 2000. Reported outbreaks included legionellosis, typhoid fever, salmonellosis, viral gastroenteritis, enterotoxigenic E coli infection, shigellosis, cryptosporidiosis and trichinosis (WHO 2004). Pathogens can be introduced via viremic passengers and crew, transported livestock, produce, and cargo. Containerized cargo greatly enhances the probability of the introduction of pathogens (Reiter 2010). Recently, epidemic diseases such as West Nile Virus and St. Louis Encephalitis have been introduced to the United States. These viruses are transmitted to humans by infected arthropods and are called arboviruses. In each case it is not exactly clear if the disease was introduced by viremic individuals infecting local arthropod populations or by transmission to humans from infected arthropods in infesting cargos or baggage. Dengue and Malaria are similarly transmitted arboviruses and there is a continuing threat that they will be introduced by shipping particularly from underdeveloped country. The Port of Houston as do other international seaports has the challenge of preventing the introduction of infectious disease through shipping without creating disruptions or barriers to the free flow of commerce. Mathematical models (both discrete and continuous) can be employed to predict the circulation and spread of a disease through a given population or populations in a given geographic subsequent to its outbreak and to help plan strategies for public health interventions (Reiter 2010). However, we need to develop models which address the global transmission of viruses.

2.5.5 *Oil Spills*

Given the extensive refining and petrochemical activity the Houston Ship Channel and the Port of Houston, there is always concern about oil spills. Oil spills have the potential to shut down the channel for protracted periods of time thus they not only can have a devastating environment effect and damage the local community they can create severe economic damage. The environmental damage is often long term because the principal components of crude oil, polycyclic aromatic hydrocarbons, can last for years in sediment and in marine environment and can be very difficult to clean up. The rupture of transporting tankers as a result of collisions with other vessels (typically ships or barges) or with infrastructure or running aground is a major and the most dramatic source of large spills. Such occurrences are caused by people and cannot be prevented altogether. However there are ways to mitigate the risk that include: improving the design of ships with double hulls and smaller individual tanks, enhanced control and monitoring of vessel traffic, strict enforcement of regulations, and insuring the proper training of crews. A more common albeit less dramatic and less visible cause of maritime oil pollution is what one might call operational. Operational spills are caused by human error or malfeasance. For example operational spills can result from the improper handling of bilge water or during the cleaning of tankers after offloading (Schmitz-Felten 2016).

The computational modeling of oil spills has been an active area of research over the past twenty five years and many issues are yet to be resolved (Reed et al. 1999). Computational models can be used to predict the evolution and behavior of an oil slick and they can be an important tool for managing the response after a spill. Models range from simple particle models to complex multi-physics systems that couple hydrodynamic and meteorological models as well as incorporating chemical and physical process and biodegradation (do Carmo et al. 2010). In order to be useful in assessment of the implementation of containment measures simulations of numerical model should be sufficiently fast to incorporate real-time information. The models must be sufficiently robust because in real situations there is uncertainty in both the quantity of oil spilt and some of its material properties (Reed et al. 1999).

References

— (2010) Numerical models of storm surge, wave, and coastal flooding

Al-Hamad K, Al-Ibrahim M, Al-Enezy E (2012) A genetic algorithm for ship routing and scheduling problem with time window. Am J Oper Res 2:417–429

AMEPA, Australian Marine Environmental Protection Association (2008) Main causes of oil pollution by ships

American Petroleum Institute (2003) Security vulnerability assessment for the petroleum and petrochemical industries, Washington DC

Bakir NO (2007) A brief analysis of threats and vulnerabilities in the maritime domain. In: Managing critical infrastructure risks. Springer, Heidelberg, pp 17–49

Balduzzi M, Pasta A, Wilhoit K (2014) A security evaluation of AIS automated identification system. In: Proceedings of the 30th annual computer security applications conference. ACM, pp 436–445

Bethel HL, Sexton K, Linder S, Delclos G, Stock T, Abramson S, Bondy M, Fraser M, Ward J (2006) A closer look at air pollution in houston: report of the mayors task force on the health effects of air pollution. Institute for Health Policy University of Texas School of Public Health, Technical Report

Birwas SK (2000) Technical aspects of hazards management in refineries and petrochemical plants. In: Petroleum refining and petrochemical based industries in Eastern India. Allied Publishers, New Delhi, pp 43–52

Blackburn JB, Bedient PG, Dunbar LB (2014) SSPEED Center 2014 Report. Technical report SSPEED Center, Rice University, Technical Report, 56 p

Board Transportation Research (2003) Shipboard automatic identification system displays: meeting the needs of mariners. National Research Council (US), Committee for evaluating shipboard display of automated identification systems

Bronk C (2014) Hacks on Gas: Energy, Cybersecurity, and US Defense. Baker III Institute for Public Policy, James A

Burns MG (2014) Port management and operations. CRC Press, Boca Raton

Caldera EJ, Ross KG, DeHeer CJ, Shoemaker DDW (2008) Putative native source of the invasive fire ant Solenopsis invicta in the USA. Biol Invasions 10(8):1457–1479

Christiansen M, Fagerholt K, Nygreen B, Ronen D (2007) Maritime transportation. Handb Oper Res Manag Sci 14:189–284

Coelho LC, Cordeau J-F, Laporte G (2013) Thirty years of inventory routing. Transp Sci 48(1):1–19

do Carmo JA, Pinho JL, Vieira JP (2010) Oil spills in coastal zones: predicting slick transport and weathering processes. J Open Ocean Eng J 3: 129–142

Fitzgibbon WE, Martyenko A (2002) Mathematical analysis of atmospheric pollution models. Technical Report

Hwang, S-J (2005) Inventory constrained maritime routing and scheduling for multi-commodity liquid bulk

Khan AA (1999) Risk assessment in the petrochemical industry. In: First international seminar (SAFE 99) in safety and fire engineering

Kramek CJ (2013) The critical infrastructure gap: US port facilities and cyber vulnerabilities. Federal Executive Series Policy Papers, Brookings Institution

Lea CJ, Ledin HS (2002) A review of the state-of-the-art in gas explosion modelling. Health and Safety Laboratory

Lewis S (2006) SHEPERDs delight: a new tool for explosion exceedance

Miller AW, Minton MS, Ruiz GM (2011) Geographic limitations and regional differences in ships ballast water management to reduce marine invasions in the contiguous United States. Bioscience 61(11):880–887

Ng WC, Mak KL (2005) Yard crane scheduling in port container terminals. Appl Math Modell 29(3):263–276

Obama B (2013) 2013 State of the union address. Speech given at U.S. Capitol, Washington, DC, February 4, 2013

Parfomak PW, Frittelli J (2007) Maritime security: potential terrorist attacks and protection priorities. DTIC Document

Persson JA, Göthe-Lundgren M (2005) Shipment planning at oil refineries using column generation and valid inequalities. Eur J Oper Res 163(3):631–652

Port of Houston Authority (2016) About the Port

Reed M, Johansen Ø, Brandvik PJ, Daling P, Lewis A, Fiocco R, Mackay D, Prentki R (1999) Oil spill modeling towards the close of the 20th century: overview of the state of the art. Spill Sci Technol Bull 5(1):3–16

Reiter P (2010) The standardised freight container: vector of vectors and vector-borne diseases. Revue Scientifique et Technique (International Office of Epizootics) 29(1):57–64

Rivera C, Mellqvist J, Samuelsson J, Lefer B, Alvarez S, Patel MR (2014) Quantification of NO2 and SO2 emissions from the houston ship channel and Texas city industrial areas during the 2006 Texas air quality study. J Geophys Res Atmos

Schmitz DC, Simberloff D (2001) Needed: a national center for biological invasions. Issues Sci Technol 17(4):57–62

Schmitz-Felten E (2016) From OSHwiki

Seebens H, Gastner MT, Blasius B (2013) The risk of marine bioinvasion caused by global shipping. Ecol Lett 16(6):782–790

Stockwell WR, Lawson CV, Saunders E, Goliff WS (2011) A review of tropospheric atmospheric chemistry and gas-phase chemical mechanisms for air quality modeling. Atmosphere 3(1):1–32

Takahashi K, Watanabe K (2010) Advanced numerical simulation of gas explosions for assessing the safety of oil and gas plant. INTECH Open Access Publisher

WHO (2004) Ships and public health. Intergovernmental working group on the revision of the international health: regulations. Working Paper 12 (12 January 2003)

Yang W, Deng L, Niu Q, Fei M (2013) Improved shuffled frog leaping algorithm for solving multi-aisle automated warehouse scheduling optimization. In: Asian Simulation Conference. Springer, Heidelberg, pp 82–92

Chapter 3
Agile Deep Learning UAVs Operating in Smart Spaces: Collective Intelligence Versus "Mission-Impossible"

Michael Cochez, Jacques Periaux, Vagan Terziyan and Tero Tuovinen

Abstract The environments, in which we all live, are known to be complex and unpredictable. The complete discovery of these environments aiming to take full control over them is a "mission-impossible", however, still in our common agenda. People intend to make their living spaces smarter utilizing innovations from the Internet of Things and Artificial Intelligence. Unmanned aerial vehicles (UAVs) as very dynamic, autonomous and intelligent things capable to discover and control large areas are becoming important "inhabitants" within existing and future smart cities. Our concern in this paper is to challenge the potential of UAVs in situations, which are evolving fast in a way unseen before, e.g., emergency situations. To address such challenges, UAVs have to be "intelligent" enough to be capable to autonomously and in near real-time evaluate the situation and its dynamics. Then, they have to discover their own missions and set-up suitable own configurations to perform it. This configuration is the result of flexible plans which are created in mutual collaboration. Finally, the UAVs execute the plans and learn from the new experiences for future reuse. However, if to take into account also the Big Data challenge, which is naturally

M. Cochez (✉) · J. Periaux · V. Terziyan · T. Tuovinen
Faculty of Information Technology, University of Jyväskylä,
P.O. Box 35 (Agora), 40014 Jyväskylä, Finland
e-mail: michael.cochez@fit.fraunhofer.de

J. Periaux
e-mail: jperiaux@gmail.com

V. Terziyan
e-mail: vagan.terziyan@jyu.fi

T. Tuovinen
e-mail: tero.tuovinen@jyu.fi

M. Cochez
Fraunhofer Institute for Applied Information Technology FIT,
53754 Sankt Augustin, Germany

M. Cochez
RWTH Aachen University, Informatik 5, 52056 Aachen, Germany

J. Periaux
International Center for Numerical Methods in Engineering (CIMNE),
08860 Castelldefels, Barcelona, Spain

P. Diez et al. (eds.), *Computational Methods and Models for Transport*,
Computational Methods in Applied Sciences 45,
DOI 10.1007/978-3-319-54490-8_3

associated with the smart cities, UAVs must be also "wise" in a sense that the process of making autonomous and responsible real-time decisions must include continuous search for a compromise between efficiency (acceptable time frame to get the decision and reasonable resources spent for that) and effectiveness (processing as much of important input information as possible and to improve the quality of the decisions). To address such a "skill" we propose to perform the required computations using Cloud Computing enhanced with Semantic Web technologies and potential tools ("agile" deep learning) for compromising, such as, e.g., focusing, filtering, forgetting, contextualizing, compressing and connecting.

3.1 Introduction

What is a Smart City? From the time of first buildings made by humans up to this day, we have considered buildings as being static and stable structures, which provide local functionality, like shelter and warmth. They were constructed with a given purpose which, of course, could change slowly over time. However, the buildings in a city did themselves not have any direct interactions with—or understanding of—each other. The revolution of information technology has changed this set-up dramatically. Today we can discuss about environments which have a consciousness, i.e., building and structures with some level of understanding about themselves and others. Obviously, this understanding is artificial and based on algorithms developed by programmers, but will anyway change our conception about the place we life today.

Smart cities can be identified on many levels. The identification can be close to daily life, like a smart environment, mobility, and living, or more abstract like a smart economy and governance or even smart people. The practical focus of recent studies are the growing amount of information and data and how to analyze and use it. We regard a smart city as a large multilevel construction where the buildings, streets, shops, bridges, traffic lights and so on, can discuss, collaborate, make decisions, and perform actions with or without human intervention. Using conventional technology, like e.g. the Internet, it allows to objects a possibility to share and gain information from their environment. This knowledge can be harvested by using multiple sensors, like, for instance, cameras and thermometers. Combining and analyzing this jointly collected information could potentially enrich the sense of the situation for each individual object and allow more accurate responding. All in all, this could provide more cost-effective solutions and real-time services for humankind.

When we also consider moving sensors and actuators, like unmanned aerial vehicles (UAVs), the amount of collected data and possible actions will increase dramatically. Earlier, discussions were focused on static objects gaining capabilities by having some form of consciousness. Next, we can think about what the opportunities are when these objects would gain the ability to move. It is a known fact, that static sensors and cameras suffer from their limited degrees of freedom. Further, special measurement equipment is, even today, still quite expensive and there is no way to insert them everywhere where needed. For these reasons, in large areas, there is no possibility to cover the area densely enough for reliable automatic decision making. Therefore, there is an evident benefit when employing devices which carry the sensors around, like UAVs or fleet of UAV's.

Worldwide there is a growing interest in the use of mobile robots and different types of moving sensors. Popular examples of applications of that type of devices are found not only in homeland security, border protection, and vessel tracking but also in logistics and crisis management (fires, accidents, floods...). Perhaps the most versatile type of devices are UAVs. These devices, sometimes called drones, have a high degree of mobility, can carry a wide range of sensors, and are able to perform a wide range of actions in the environment using actuators.

In this paper, we discuss aspects of using a fleet of collaborative unmanned aerial vehicles to complete missions. Especially, we consider a possible strategy at how these devices can be employed during emergency situations in smart cities and how Cloud Computing technologies can be used to aid their working. The idea is to inspire the audience to take actual steps for practical implementation of the proposed automated services. This article is an extension of previous work, published in Cochez et al. (2014).

For illustrative purposes we will use the following scenario:

- There is a big fire in the forests close to a major city. Firemen are fighting the fire using helicopters and specialized UAVs. Further, cameras placed on other UAVs are used to observe the area. The observations of the UAVs are collected in a server which is hosted in the Cloud.
- At some point it starts to rain heavily and thanks to the water the fire gets extinguished.
- There is, however, so much rainfall that the city is threatened by it. However, based on the readings from sensors placed in the rivers and observations made by UAVs, the people in the city are warned in time.

In the next Sect. 3.2, we will give a broader description of the problem. Then we describe how the actions needed in emergency situations can be regarded as an optimization problem in Sect. 3.3. In Sect. 3.4 we will discuss the potential of using Cloud Computing as a resource of computational power. Next, in Sect. 3.5 we discuss how we use agents, which use a semantic model for reasoning and data storage, to help the UAVs to perform their tasks. The amount of data collected by the UAVs is of such volume, arrives at such a velocity and is of such nature that we can consider it a Big Data problem for which we provide a solution in Sect. 3.6. In the last Sect. 3.7 we will conclude and describe future research directions.

3.2 Emergency Management in Smart Cities

In this section, we will focus on one challenging application area, namely, emergency management in smart cities. These situations, which usually evolve in an unpredictable manner, will require real-time optimization, and the deployment and coordination of a temporary, adaptive fleet of intelligent UAVs. There is also a need for a strategy which optimizes the Cloud-driven data collection and the distribution of data between populations, ground support and control centers.

Past studies have been mainly focused on the cases where there is only one vehicle without any dynamic cooperation, see for instance (Gonzalez et al. 2009). In some cases cooperation has been implemented using pre-planned routes and tasks. In the paper by Doherty et al. (2013), the authors acknowledge that in order to increase autonomy in heterogeneous UAV systems, there is a need for automatic generation, execution and partial verification of mission plans. We would like to augment this statement by saying that also validation of the plans is needed, i.e., it is not enough if the plan complies with a given specification, it must also be assured that the plan meet the needs of all stakeholders.

The robots or drones can use each others' observations to decide upon their own actions and hence disable some of their own sensors in order to save energy. Furthermore, a UAV squadron can be seen as a toolbox with several tools, where one can be used for the collection of information, the other will encrypt everything for secure communication, one can provide specific information for optimization calculations and yet another one can ensure communication with the Cloud. In our example scenario we have UAVs, which are specifically designed for extinguishing fire; this group of UAVs has specific physical properties, which make them a perfect tool for the given task.

UAVs could also be used for preventive observations. Therefore, it is desired that a UAV can be instructed to limit how disturbing it is. This can be done by lowering its visibility and noise level, as opposed to the loud and visible activities which are likely during fast operation. In another situation, there could be a 'mother ship' with reloading capabilities and the ability to fly long distances. The number of options is enormous and the choice depends on the mission at hand.

The dynamic nature of this problem makes this system most interesting. There are plenty of possibilities for failure and a human programmer cannot take each and every aspect into account before the mission, especially because the mission will evolve over time. Actually, the kind of situations which we are talking about are very much like the 'Black Swan' as described by Taleb (2010), i.e., an event which lies outside the realm of regular expectations, has a high impact, and is retrospectively predictable. The predictability of the situation can be disputed, but at least the first two properties imply that it is not feasible to create static plans to cope with the situation. In our running example, a static plan could perhaps not have predicted that the rainfall would be so heavy that it would flood the city.

Missions should not fail if one UAV breaks or gets inhibited, nor if network communication does not complete as assumed. Therefore, we need a hierarchical model for different kinds of situations. This model depends on the additional information available during missions. At the low levels, we do not consider collaboration at all, the idea is that each UAV survives the mission as a 'lonely wolf'. At higher levels there will be an increase in the amount of information and collaboration available and more computational intelligence can be used in the mission control. At the highest level, the squadron works as a one big creature which limb are connected using information bonds. During the mission, we have to dynamically move up and down in this hierarchy and, at the same time, single UAVs have to learn how to survive and accomplish their mission. Clearly, one of the main questions is how to assign tasks and resources while controlling this complex system. From our example, one could imagine what happens when a UAV full of water looses the connection with other UAVs and the central control system. It should most likely take the autonomous decision to drop the water of at the location with most fire which it can observe and get back to get more water.

Operating multiple UAVs gives the possibility of fusing their combined capabilities, potentially leading to rich insights regarding the environment. However, combining the data to generate new knowledge and making the UAVs to react accordingly in a coordinated fashion leads to a high computational overhead. It is clear that the processing power which can be carried on board of the drone will not be adequate for making optimal coordinated decisions. Further, events registered by the UAVs or inferred from analysis of the collected data will lead to bursts of high near real-time computational need. The UAVs should collect as many data as needed about the situation which need to be analyzed to infer the actions which should be taken. Most likely in collaboration with human experts. In our example, UAVs should collect data from the fire situation using, for instance, infrared cameras. These measurements are then collected and a system should filter out the most interesting ones for human consideration. It is important to also keep a record of past observations, as they can, when combined with maps of the area and weather information, be used to make a prognosis on where the fire will move to. Calculating these prognoses is too computationally intensive for UAVs, already assuming that they would have all the other needed data available. It is also clear that these calculations are only needed in exceptional situation.

We propose that Cloud Computing is the most suitable option to solve the issue of bursts in processing power needs by providing adaptive computational power and storage capacity. Usage of Cloud Computing also improves fault and failure tolerance by providing the possibility to globally replicate large sets of data and distribute work load. Cloud Computing will, however, have an unacceptably large latency for certain tasks which will have to be solved on the UAVs themselves. As a consequence, there is a need for an adaptive hybrid solution with Cloud Computing and the real-time systems on board (see abstract architecture in the Fig. 3.1). We will discuss this issue in more detail in Sect. 3.4.

3.3 Why to Optimize?

The overall goal of the system is to keep itself in an ideal configuration for all missions it should perform simultaneously, given the evolving operational environment. A fleet of many UAVs and other autonomic devices can have numerous states. Factors, which are part of the overall state, include the environment, the amount, type, location and configuration of devices, the state of the mobile network used for communication,

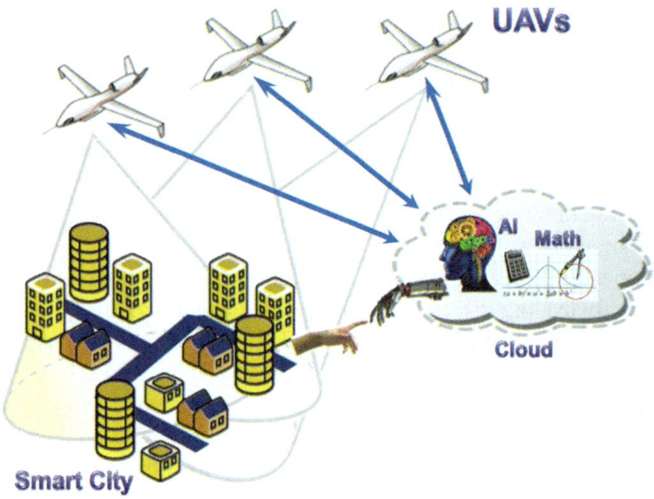

Fig. 3.1 Abstract architecture of the system

and the computational load and data stored on the devices. All these factors have an influence on the performance of the system. When the performance of the system is at its maximum, for that particular moment, the system has an ideal configuration. Clearly, the problem of finding an optimal configuration at each point of time leads to a computation so complex that it becomes unfeasible to perform, this can be seen easily since it is already impossible to model for instance a single UAV 100% accurately. The number of influencing factors is of such a magnitude that it is comparable to a situation which requires reasoning using the open world assumption (see Sect. 3.5) and we will have to be satisfied with models and approximations and hence quasi-optimal solutions for the problem.

The flying routes are one aspect which can be considered for optimization. During an emergency situation the best outcome is, in most cases, the minimum time needed for reaching the target location. However, we also need to make sure that the droid does not fly into areas which are dangerous like, for instance in our running example, a fire. Determining an optimal solution can be reduced to finding a minimum of a given cost function defined for the mission. However, in practice, the cost function will be too complex (e.g., not derivable), too expensive and too dynamic to solve analytically and we will have to consider the problem in a more pragmatic way. One promising approach is the use of some form of evolutionary computing since it allows for the function to be considered a black-box, i.e., an analytical or functional description of the function is not required. Instead, it suffices that we are able to compute values from the function. Further, the parameters for some of these algorithms can be tuned such that solving a dynamically changing problem becomes possible. Lastly, most of the algorithms can be highly parallelized as was, for instance, done by Fok et al. (2007).

Another aspect which can be optimized, is the communication environment. The missions should, if possible, be planned such that the network remains reachable. Promising work has been performed in the field of Wireless Mesh Networks. This approach enlarges the coverage of the network and should help the devices to save power, which is obvious bottleneck of the system. A smart choice in routing paths and link scheduling algorithms can further lower interference, balance the load, and enhance the throughput and fairness as was studied by Pathak and Dutta (2011). It should also be observed that the situation is quite different from traditional wireless mesh networks because the nodes work cooperatively. We could, for instance, instruct one of the robots to move to a certain location where a higher transfer rate can be obtained, i.e., we can physically move the medium containing the data and even connect the device to a wired link in case of need. In the case of a fire we could for instance have one, perhaps physically specialized, droid which places itself between the fire zone and the central control station or closest radio mast. This droid could then relay the communication traffic from the others and hence reduce the transmission power needed, which also improves the transfer rates. Kopeikin et al. (2013) used agents to simulate a more complicated setting where the choice and location of UAVs to be used as relays is depending on their utilization rate and network topology prediction. We select their approach for our setting since keeping the communication link working is, despite the fact that we plan for a system which can work without central control, a much desired property.

In the setting we are describing we will have to find optimal routes, communication, and other factors, which are contradicting objectives (see e.g. conflicting games in Nash 1951). Hence, we have to consider the case as a multi-objective optimization problem and will need to use appropriate methods for solving it. It also turns out that optimal becomes somewhat ill-defined since there will be multiple solutions which can be considered optimal, i.e., there will be a Pareto front of optimal solutions in the case of cooperative games. Using Cloud Computing we might, as we elaborate further in the next section, be able to consider all factors for the optimization of the multiple tasks which the system has to deal with simultaneously. On the practical level, a hybrid optimization environment has to be created. This environment should consider many algorithms (coalition games) for solving problems and simulate their outcomes before deploying the best ones to the UAV. Also the method used for finding out the best algorithm is likely to evolve over time, we develop this idea further in Sect. 3.6.

3.4 Computing Resources

In this section we will look at how Cloud Computing can help us in emergency situations with the help of UAVs and other devices with a degree of mobility. We will discuss why a dynamic combination of Cloud Computing and the computational power of the components in the network should be used in an emergency situation, how the Cloud can help to handle bursts in the need for computation, how interaction with human gets facilitated, and discuss a model for data storage which we elaborate further on in the next sections.

In their article (Mell and Grance 2009) defined Cloud Computing as

> a model for enabling ubiquitous, convenient, on-demand network access to a shared pool of configurable computing resources that can be rapidly provisioned and released with minimal management effort or service provider interaction.

In their studies, (Buyya et al. 2009), investigate Cloud Computing as an emerging IT platform and compared it to other technologies like grid and cluster computing. They identified the differences in resource requirements of scientific, data analysis and Internet based applications. In emergency cases, the Cloud will be used from all three perspectives. Scientific computations and data analysis are needed for the analysis and the verification of models and strategies. Much of the interaction with the Cloud will happen trough Internet based applications, since it is the ubiquitous way of interacting with Cloud Services. The most important reasons why we select Cloud Computing as a solution for our scenario are rapid scalability (both up and down), high security and privacy, and strong support for fail-over and content replication. These features are not all at the same time available in either grid or cluster computing. We expect the services which are located in Cloud Computing infrastructure to assist the overall system by coordinating and (re)configuring the other components.

Research on changing transport situations has some similarities with coping with disasters: in both scenarios there is an amount of slowly evolving infrastructure, e.g., the roads and the cities' shape, faster and repetitive events, e.g., the behavior of road users and city inhabitants, and unexpected situations with a severe impact, e.g., traffic accidents and wide scale disasters. Earlier research by Li et al. (2011) proposes the use of Cloud Computing for solving simulation and decision making problems for urban transportation systems. In their research the authors indicate that for performing traffic simulations both great amounts of computing power and storage capacity are needed. These simulations happen both on-line and many of them can be ran in parallel.

They conclude that Cloud Computing is a reasonable approach since a supercomputer would imply a risk for insufficient capacity in the future. A supercomputer would also suffer from a lot of wasted capacity while in the Cloud Computing setting multiple systems can share the Cloud's "infinite" capability. In a similar realm we want to tap into this capacity to solve hard problems in a short time.

A system can use several strategies to support the missions. The basic options can be found in Trigui et al. (2012) which distinguishes a centralized strategy, a distributed strategy, and a market-based approach which is a trade-off between these two.

The centralized strategy, where one central component has control over the whole system, sounds like a reasonable solution when combining the multiple components with the help of a central Cloud Computing based solution. This strategy was also most time efficient in the simulations by Trigui et al. (2012) where a combination of mobile robots and a Wireless Sensor Network were used to optimize the capturing of intruders. However, it is noted in the paper that

> [the simulation] did not take into consideration the communication delays and we provide insights only on the pure behavior of the coordination strategies independently from the communication.

In the scenario which we consider these delays in communication can have an unacceptably large latency for certain tasks. For instance, a robot which detects an obstacle using an on-board camera cannot send the data to a central place and wait for a rerouting. The navigation decisions should be performed autonomously as, for instance, demonstrated by Kang and Prasad (2013) for an uncertain environment. Other tasks, such as moving object recognition (Nimmagadda et al. 2010), can be solved partially on the UAV or completely on a centralized server, dependent on factors like wireless bandwidth, image complexity, and the size of the reference image database. In their particular experiment, it was always more favorable to offload the task partially or completely to the server, when the image database contained 5000 images or more. The authors also use a speed-up factor to indicate the computation speed difference between the centralized system and the UAVs, and note that this factor can be treated infinity in the case of Cloud Computing.

Another argument against the use of a centralized system is that it becomes a single-point-of-failure. This last drawback is mitigated by the use of Cloud Computing since, as discussed by Buyya et al. (2009), it has a strong support for fail-over and content replication as compared to grid or cluster computing which usually only support restarting failed computations.

In the case of an emergency it is very well possible that the central control station becomes unreachable. When in this scenario the centralized strategy is implemented, the robots would become unusable since they are totally dependent on orders from the central control. The distributed strategy seems to offer a solution; each robot or agent is making a decision based on information which is available locally. However, as shown in the experiments by Trigui et al. (2012), the distributed strategy suffers from an increased mission time. This is as expected since the individual components will rarely be able to compute a global optimal solution by only using the locally available information. A clear benefit is, however, that if the volume of data needed to make the centralized decision becomes huge or the communication is unreliable or slow, the mission time for a distributed system would actually become lower.

The final strategy considered is the marked-based approach which does not require all information to be sent to the central control station. Instead, as argued by Dias et al. (2006), to perform a given task the robots should be chosen based on their own estimation of resource usage. In other words, the robots make their own estimation of the cost and based on this figure a central decision making entity decides which robot will perform the job. The measurements in Trigui et al. (2012) show that in terms of mission time, this strategy performs somewhere in between the centralized and the distributed strategy. The distance which each robot travels, i.e., the resource use, is not significantly different from the centralized strategy.

As argued above, both a purely centralized or distributed setting have undesired properties when using multiple UAVs for emergency situations. These factors are quite specific to the setting. When using a centralized setting, like a purely Cloud based solution, the system will have a latency which is unacceptable for flying robots. A purely distributed setting would then again have the drawback that a global optimum is most likely unattainable. The consequence of these two extremes, complete autonomy of the UAVs and complete offloading to the Cloud, shows the need for an adaptive hybrid solution with Cloud Computing and the real-time systems on board (see abstract architecture in the Fig. 3.1). In other words, we will need something close to the market-based approach which combines the centralized and the distributed strategy. This dynamic strategy would continuously evaluate which of the strategies is most beneficial at a given time-point and would also have to take the survivability of the system into account. The robots will then in case when the central control station cannot be reached for a certain amount of time continue to work autonomously, perhaps with a lower performance. To attain this, the functions of the Cloud which are essential for the working of the overall system have to be replicated in the UAV network. At any point, the Cloud should simulate and try to predict what would happen in case of different kinds of failures and pro-actively adapt the algorithms on the UAVs.

It is expected that certain tasks which we would like the Cloud to solve are not computable within reasonable time, or not at all, due to an inherent complexity of the problem at hand. Hence, the Cloud needs to hand certain tasks over to humans in order to get them solved. Further, interactions with humans are also needed to assist rescue teams in case of disasters. We expect that a certain subset of the tasks will be of such nature that they can be solved using a technique similar to Amazon's mechanical Turk (see https://www.mturk.com/). This crowd-sourcing platform enables its customers to create *Human Intelligence Tasks* and decide upon the price which human workers will get paid upon completion of the task. This kind of tasks can be, for instance, the annotation of imagery as was studies by Sorokin and Forsyth (2008). The authors concluded that

> the annotation happens quickly, cheaply and with minimum participation of the researchers, we can allow for multiple runs of annotation to iteratively refine the precise definition of annotation protocols.

One important concern with the use of crowd-sourcing is that in case of a disaster related to a conflict between humans, the opposite party might try to counterfeit the answers to the system. In the case of our illustrative example, we could use a Human Intelligence Task to check preselected frames, recorded by the UAVs, to decide whether there are humans visible.

The Cloud and modern web infrastructure is also known for the introduction of NoSQL databases. This term refers to several types of databases which do not support SQL with traditional ACID guarantees. This abbreviation stands for Atomicity, Consistency, Isolation, and Durability, which means that any transaction in the database happen completely or not at all, will not violate database rules, happens without interference from other transactions, and is stored in the database even in case of failure immediately after the transaction, respectively, see also Haerder and Reuter (1983).

Many of these NoSQL databases have been designed to handle specific types of data or huge amounts of data. We will discuss more about a certain type of graph data, which we choose to use for storing data, and the handling of huge amounts of this kind of data in the following sections.

3.5 Use of Semantic Agents

From an abstract perspective, a Smart City can be seen as a collection of entities with some kind of identity and relations between them. When we consider this idea, we can see that we could represent the Smart City as a graph where vertices model entities and edges are used to describe their relations. This observation suggests that the Semantic Web, as originally suggested by Berners-Lee et al. (2001), would provide suitable concepts for describing and reasoning about Smart Cities. The Semantic Web encompasses the idea that entities and the relations between them in the real world can be described in a similar way as web documents and links between them. The entities of interest, in a Smart City, have some kind of functionality, and many have sensors, actuators or some sort of, perhaps artificial, intelligence. This is to say that the abilities of the entities vary greatly. Examples of entities do not only include wireless sensors for temperature and pollution, traffic and street light control systems, and pavement heating, but also digital services and the people living in the city. Also the UAVs are just examples of entities. The integration for these types of entities was studies before in Terziyan (2008) and each of them was regarded as a Smart Resource. In the same article the author also suggest that the Semantic Web does not reach its aims if it can only be used to describe passive, non-functional entities. Instead, it should be possible to describe the behavior of entities, which includes communication, coordination, and proactivity, i.e., the ability to predict the upcoming situation and act accordingly. The suggested approach was also further enhanced in Katasonov et al. (2008), which adds semantic agents to the picture. These agents are a representation of the Smart Resource in an agent environment. By encapsulating each Smart Resource with an agent, the entities gain the ability to communicate with each other. When this communication happens in a Semantic language which allows for dynamics, like, for instance, S-APL formalized in Cochez (2012), we can reach the goal of describing the behavior of entities.

When looking at disasters in Smart Cities, it appears reasonable, or perhaps even natural, to use Semantic Web concepts to describe and exchange information. The reason is that disasters tend to happen unexpectedly, and when they happen they have a great local impact. Also the time-scale of disasters varies a lot, the flow of information is raising over time, while available resources, like batteries, decrease. This unexpected nature implies that it is hard, if not impossible to give a framework or fixed set of terms to describe a situation. Further, we are not able to identify beforehand all entities which would play a role during the development of events. Therefore, we need a way of describing any entity and a way of reasoning about them. This flexibility is offered by the Open World assumption, which is one of the cornerstones of the Semantic Web. This assumption encompasses the idea that if we do not know something about an entity we cannot assume that it is not true, i.e., we always assume that our graph of data is incomplete. If we, from our illustrative scenario, have not received any information about the flood yet, we can still not make the assumption that everything is fine. It could namely be that certain UAVs are saving their energy by not communicating immediately. This view and way of reasoning is opposed to a Closed World assumption, where the fact that something is not known would immediately imply that it is not true, i.e., the world exists only of the part which we are able to observe. The closed world way of reasoning is typically used in relational databases.

The main drawback of using Semantic technologies could be the reduced processing speed. Effort related to making semantic data storages better able to cope with big amount of data can, for instance, be found in Abadi et al. (2007). We, however, expect that not the storage, but the processing might become the bottleneck in a concrete implementation, partially caused by a need for alignment and negotiations. This extra processing power could be provided by the Cloud infrastructure at the cost of latency as described in Sect. 3.4.

The use of an Agent System is a natural choice since it reflects the actual environment to a great extent. The UAVs can be represented by agents because they are able to act, to perceive their environment, and to communicate with each other. Further, we are looking into UAVs which can function autonomously and which have the ability to achieve their goals. These requirements, needed for the UAV to be classified as an agent, and further research on distribution of intelligence and coordination among agents, have been collected in Ferber (1999). It should be noted that the same work also handles the concept of organizations and sub-organizations. An organization groups agents together based on their functions and roles. One of these organizations can itself be considered as an agent in a bigger organization. Similarly, when we are combining Cloud Computing into our agent system, the Cloud itself is regarded as an organization and hence also as an agent inside the whole.

Joining the capabilities of Agents and Cloud Computing has been proposed earlier in the context of Smart Cities, like, for instance, by Ito et al. (2012) who proposed their combined use for solving questions about smart housing, senior people care and road congestion management. Further research, related to the use of agents and Cloud Computing in traffic situations, can be found from Li et al. (2011). Their main argument in favor of using a multi-agent setting is that mobile agents only require a runtime environment and can run their computations near the data which increases performance, reduces latency and communication costs. Moreover, a multi-agent system is said to perform better as a static agent system when faced with requirements of dynamic traffic scenes. The agents are able to travel trough the component involved and autonomously implement strategies. It is also stated that thanks to the use of Mobile Agents new strategies can be distributed while the system is running and hence it is possible to keep up with scientific development. Based on these ideas we add the requirement for the agents to be mobile to our system. The UAVs will be mere shells in which the mobile agents reside and which offers them services.

Further research on similar ideas, but focused on transportation, by Wang and Shen (2011) points out that contemporary Cloud offerings also include infrastructure beyond general purpose computers. Especially when combined with agents which share similar behaviors it is possible to harness the power of parallel execution. The authors proposed and experimented the use of Graphics Processing Units (GPUs) as a host for traffic signal timing optimization problems. This type of hardware has also been used earlier for running evolutionary computations as, for instance, by Fok et al. (2007).

The fact that we allow the UAVs to communicate with each other trough agents and to the Cloud creates many opportunities. First, the computational power of the Cloud can be used to calculate better cooperative mission paths compared to what would be possible with the limited computational power of available on the drones. Even the combined resources on the UAVs would be dwarfed by the virtually unlimited power available in the Cloud. However, as discussed before, the result of doing the computation in the Cloud can be better, but will always involve a certain latency. And when the connection to the Cloud is lost, the distributed intelligence of the UAVs, which can be managed by the agents, should be used instead.

Second, the robots could use each others' observations to decide upon their own actions and hence disable some of their own sensors in order to save energy. This also applies to other actions like physically altering the environment and even flying. A UAV which is low on energy could forward a task to another UAV, which has plenty of resources left or there could be a carrier UAV which carries other smaller UAVs on-board to save the energy needed while flying.

Finally, loosing an individual UAV is not such a problem since the other UAVs can try to take over the task by migrating the mobile agents. This is facilitated by the Smart Resource framework because the mobile agents are build as semantic agents which are able to communicate with each other about their behavior and capabilities.

3.6 Toward Evolving Knowledge Ecosystems for Big Data Understanding

We expect the total amount of data which the all sensors in the Smart City collect jointly to be enormous and varying over time. It is possible to filter the data, but discarding parts which appear to be needed later could lower the potential performance of the system dramatically. The data also needs to be processed in a timely fashion, because if the system realizes earlier that a reconfiguration is needed it can save essential resources. Hence, we will need to store and process this voluminous data at a high velocity. As we argued in the previous section, we would benefit from the use of a semantic representation of the data since it allows for an unlimited extension and the use of the open world assumption. In conclusion, we could say that we have a Big Semantic Data problem since we have the three aspects which this type of problems consists of according to Zikopoulos et al. (2012), namely the need to work with a big volume of varying data with a high velocity.

In previous work (Ermolayev et al. 2013) we looked at how to solve a Big Semantic Data problem which consisted of handling streams of tokens arriving at high rate and altogether representing a huge amount of data. That work stayed on a very abstract level. The scenarios described in this chapter are concrete example of situations where such a system could be used. In the following subsections we will, based on the previous work, describe the balance between volume and velocity, how big data can lead to big knowledge, and how a system inspired by the mechanisms of natural evolution could provide a solution for managing this knowledge. This type of system should be implemented and deployed on Cloud infrastructure to provide scalability. We will also offer suggestions on how the system could work, based on our illustrative example.

3.6.1 Volume Versus Velocity

Volume, the size of the data, and velocity, the amount of time allowed for its processing are clearly the main factors when talking about our Big Semantic Data problem. Both of them manifest themselves when a high number of measurements, like for instance camera recordings, temperature, and humidity, have to be handled within a reasonable time. When the system extracts information or knowledge from the data it does this effectively if no important facts are overlooked, i.e., the analysis is complete. The effort is a measurement of how much time and resources the system uses to perform this extraction. We say that the utility of the extraction is high, when the results are useful. The ratio of the utility to its associated effort is interpreted as the efficiency of the system.

$$\frac{\text{utility}}{\text{total processing time}} = \text{efficiency}. \tag{3.1}$$

To illustrate these ideas, imagine that the system receives an image of a fire for processing. If the system is able to extract all possible information from it, including perhaps data not related to the fire, it is effective. In case these results are useful, we achieved a high utility. When the system is able to do this fast or with a low amount of resources it has a low effort measure. If all of these are true, we achieve a high efficiency, i.e., a lot of useful information got extracted with little effort. If, however, the computation and additional communication latency is slower than the required velocity, the utility of the result drops to zero.

Note that when the effectiveness of the system increases, the computational complexity will increase and hence there is a risk that the efficiency of the system drops. This does not necessarily happen in case the information found appears to be useful. Hence, if we would like to make a deeper analysis of the collected data, we would have a less efficient system. To cover up for this lower efficiency, we would need a higher computational power, which Cloud Computing can provide. The reason is that the Cloud infrastructure can reduce the effort (at least the time component, which is most relevant) to a minimum by using more resources. Obviously, even when using Cloud Computing the system will be bound by theoretical and practical limits. Moreover, as mentioned above, the computations which are performed in the Cloud come with an inherent communication latency.

3.6.2 Big Knowledge

Since we are trying to extract all relevant information from all sensors, which we have available in the Smart City, we start from a vast amount of data. Also, when we try to extract all knowledge from the data, we might end up with an unmanageable amount of knowledge. From that observation we identified some aspects which should be taken into account while working with Big Data. We called this approach *3F+3Co* which stands for Focusing, Filtering, and Forgetting + Contextualizing, Compressing and Connecting. It should be noted here that not all terms are novel in the sense that they have been used in different domains and interpretations, see, for example, (Dean and Caroline 2011) for a slightly different view on 3F. We gave an occasionally overlapping meaning to each of these terms in the context of Big Data analysis as follows:

Focusing is mainly concerned with the order in which the data is processed. An optimal focus will only scan the data, which is absolutely needed to come to an answer for the question, which is at hand and will hence lead to a higher efficiency. This facet will play a very significant role in the emergency system since the data is arriving continuously and hence the focus will need to go to the data which is related to the event which is currently going on. For instance, in our illustrative example, the focus should move from the fire fighting to the flood damage prevention as soon as there is an indication that the flood might cause more damage as the fire.

Filtering is ignoring anything, which is, hopefully, not of importance for immediate analysis. We use *hopefully* since deciding whether information is relevant or not can in most cases not be done with a hundred percent certainty. One way to filter is to only focus on specific features of the data, which also reduces its variety and complexity. The Smart City emergency system could for instance ignore information from sensors in the center of the city when there is a fire on its outskirts.

Forgetting is a further step from filtering where data, or knowledge derived from it, is completely removed from the system. This trashing can remove potentially valuable information. In the work, which we did around Evolutionary Knowledge Systems (see Sect. 3.6.3), we use the technique of "*forgetting before storing*". This means that there has to be reason before anything is stored at all in the knowledge base. Using this technique is sensible in case the amount of information which would need to be stored would be of such size that even nowadays' Cloud Computing solutions and data centers would be overwhelmed, if this is not the case it might be reasonable to keep all the data stored somewhere, at least for a certain amount of time.

Contextualizing comprises not only the change of sense of statements in different contexts, but also judgments, assessments, attitudes, and sentiments. There are various facets, which contribute to the context of data. Examples include the origin of the data, the tools used, and the place in which the result will be used. Conceptualization happens for instance when one measurement is given more importance as another one because it was made with a better measurement device.

Compressing stands for both lossy and lossless compression. Where lossy compression is similar to Forgetting which was discussed above. The lossless compression might be very effective because the high amount of data leads to a high probability that repetitive or periodical patterns are present.

Connecting can be done if information is added to an already existing body of data. The whole body is build incrementally. The benefit of linking the data before processing it further is that data and knowledge mining, knowledge discovery, pattern recognition, etc. can be performed more effectively and efficiently. A good example of this connecting used for building an index of the world wide web can be found in Peng and Dabek (2010).

3.6.3 Evolving Knowledge Ecosystems

The Cloud component of the emergency management system receives a lot of information. The content is likely to evolve over time because emergency situations are subject to rapid and unpredictable changes. To anticipate these changes the system should be able to change its inner working. In the chapter (Ermolayev et al. 2013) we proposed an Evolving Knowledge Ecosystem which is able to adapt to external change.

The core idea behind the Ecosystem is that

> The mechanisms of knowledge evolution are very similar to the mechanisms of biological evolution. Hence, the methods and mechanisms for the evolution of knowledge could be spotted from the ones enabling the evolution of living beings.

Starting from this idea, we derived that we could model the knowledge evolution inside the system using ideas from natural evolution. One of the core ideas behind the Evolving Knowledge Ecosystem is that, similar to the idea of natural selection proposed by Darwin (1859); knowledge, which is more fit for its environment, has a higher chance to survive as less fit knowledge. The environment here is formed by the incoming information to the system.

The Knowledge Ecosystem assumes that the information, which is delivered, is in the form of knowledge tokens. These tokens are self-contained fractions of the data, with a structure which allows integration into what we call knowledge organisms. In order to create these tokens, the streams of data should be transformed.

Natural language text can be converted into knowledge tokens using Natural Language Processing (NLP) techniques. The Evolving Knowledge Ecosystem is, however, not limited to that type of data. Any data stream can, in principle, be converted to a stream of knowledge tokens. In the case of our UAV scenario, the extraction of data from measurements might be even much more simple as the NLP example. Much more complicated is the conversion or analysis of images and video. The research concerning these is, due to its complexity, not as mature as for text analysis (see e.g. Atmosukarto et al. 2012 for an approach to action recognition.) There is also effort in integrating capabilities of text and media analysis by Damjanovic et al. (2011).

The benefit of using a Knowledge Ecosystem for managing the data is that it evolves over time and can hence adapt itself to new trends in the data. Also a lot of meta-data is kept inside the system and using these it is possible to propose beneficial algorithms for future actions. It could, for instance, remember that a certain area of forest burned down a year ago and realize that the fire does not go there this time. From these two fact, it could derive a rule about likelihood of spreading of a fire in a given location.

The Knowledge Ecosystem is also able to provide signals to the outside world. As a concrete example, it could instruct the UAV fleet to collect more information about the temperature in a given area. The Knowledge Ecosystem can further be queried for information and trends.

3.6.4 "Agile" Deep Learning for "Wisdom" Discovery

Song patented the architecture where an intelligent UAV performs data collection during a flight, communicates with a cloud-based control system to download selected control applications and mission-specific applications to the UAV memory, and upload data collected by the UAV to the cloud-based control system (Song 2016). Such architecture potentially enables the cloud to learn the decision models from the collected data, which are needed for controlling UAVs towards a specific mission, and communicate learned models back to the UAVs for the execution (i.e., for making real-time decisions based on the observations). The core of a traditional machine learning process is shown in Fig. 3.2. The data collected from the observations of the environment is processed by some machine learning module, which utilizes data mining and knowledge discovery instruments (see, e.g., Fayyad et al. 1996) to automatically or semi-automatically learn a decision (diagnostics, classification, prediction, recognition, etc.) model, which is assumed to be used for making further autonomous decisions on the basis of new input data. The model learning process is known to be a resource consuming task especially when dealing with large set of multidimensional data, while the model execution can be done relatively fast.

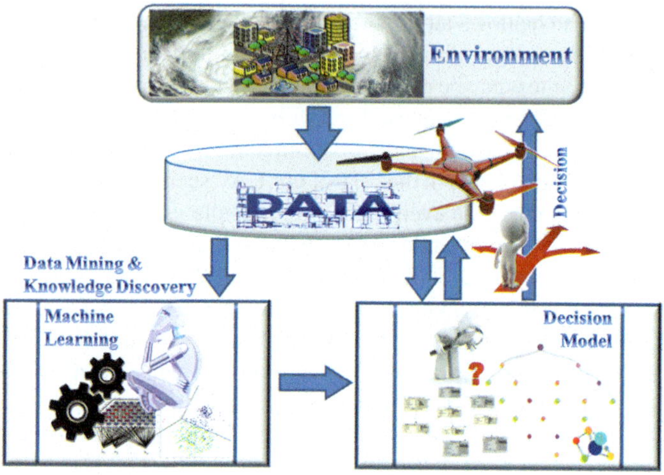

Fig. 3.2 Simplified schema of traditional machine learning application

Deep learning (LeCun et al. 2015) is needed when understanding the meaning of raw sensory input data and learning the model or function, which maps it directly to the decision, is very complicated. Deep learning resolves this difficulty by breaking the desired complicated mapping into a series of nested simple mappings, each described by a different layer of the model. The input (visible layer) contains the variables from the original observation. Then a series of hidden layers extracts evolving abstractions from the input; and the model must determine, which of these abstractions are important features (decisive, predictive, …) for the decision function (Goodfellow et al. 2016). In a recent press release Neurala, which is a Boston-based startup specialized in deep learning and neural network software for robots and drones, announced a deep learning based software solution which allows manufacturers of UAVs to install "neural" software directly into their drones for autonomous flight, object recognition, visual following and visual inspection capabilities (see e.g. Matus 2016). Machine learning in general and deep learning in particular can be a long process when addressing the big data challenge. If a situation will require having at least some (even if weak) consistent decision model somewhere in the middle of the long learning process over big volumes of data, it will not be possible as the abstractions from the hidden layers may not yet be mapped to the target attributes of the potential decisions. We believe that a good compromise between the big data challenge and the deep learning process suitable for the intelligent UAVs would be an "agile" machine learning architecture (see Fig. 3.3). Such architecture partly inherits the multilayered structure from the deep learning as well as "nesting" feature of the layers. However the philosophy of nesting will be different, i.e., agility will be the major target. We expect that the learning engine will fast built some simple (relatively "weak") model (first layer) using a restricted set of training data and only some of its features, then while the created model is making real time decisions, a more advanced ("stronger") decision model for the second layer will be created on the basis of bigger volume and dimensionality of training data, and so on. The expected benefit of the agility is sustainable decision capability of an intelligent entity as at every stage of the learning process it has some valid models capable to make decisions while the learning engine continues building more advanced decision models for the next layer. The key challenge for such architecture will be the need for very smart focusing and filtering at the beginning of the process and between the layers. The focus (first sample of training data) and selected features of it (result of filtering) is very important to guarantee at least reasonable quality versus reasonable time and memory spent for learning of the first decision model. Another interesting feature of the architecture is that the model at every layer is performing two decision tasks simultaneously:

(a) addressing real-time input and generation decisions needed for the operation of the intelligent entity;
(b) using real-time input as a context for the decision, which data points from the collected training data to "recruit" for the next expanding training sample and which data dimensions to add to the consideration. We also believe that the capabilities to filter and focus (i.e., to compromise wisely) can be learned from data in a similar way with the other decision capabilities. Such (meta-)learning skills are very important for the self-management and therefore we suggest to

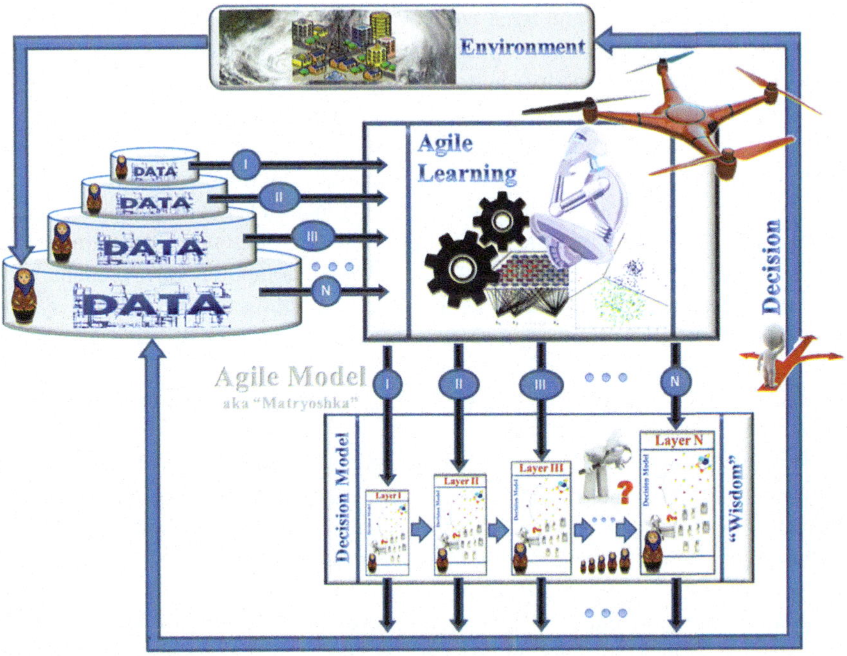

Fig. 3.3 "Agile" machine learning process generic architecture

name the agile machine learning process as the "Big Data Mining and Wisdom Discovery" opposite to traditional machine learning (aiming data mining and knowledge discovery).

The agile learning process, as presented in the Fig. 3.3, is expected to efficiently address the challenge of big data in machine learning but with a naïve assumption that the attributes of the nested data samples can be easily mapped to the target attributes or to the decision. If this is not the case at a large extend, then the best solution we believe would be a hybrid of the deep learning and agile learning architectures: Agile Deep Learning. We envision this to look like nested loops: agile—external, deep—internal, i.e., the decision model at each layer of the architecture from Fig. 3.3 is actually a multilayered deep learning structure. Still the advantages of both architectures are preserved: agility is achieved as the models are created and come into real-time operation faster and then evolve based on growing samples; quality of decisions is guaranteed by the deep learning, which computes relevant implicit data attributes at its hidden layers.

3.7 Conclusions and Future Work

In this chapter we proposed approaches related to several aspects of the use of a fleet of collaborative UAVs and Cloud Computing in Smart Cities. We started by describing what a Smart City is and what we would need from this system in emergency situations. These situations have specific characteristics like an unpredictable development and a need for immediate action. Then, we gave several examples on how the collaboration between UAVs might be used to improve the overall performance of the system. We looked at this from an optimization perspective and concluded that in order to keep the system in an optimal state, we have to continuously evaluate a multi-objective function. It was also indicated that this function is not available in analytic form and we hence proposed the use of some form of evolutionary computing which would also be very suitable for execution in a Cloud Computing environment.

Cloud Computing was selected as the main computing resource for the system since it has virtually unlimited capacity and also allows for fail-over and content replication, even geographically. Then we stated that it has to remain possible for the fleet of UAVs to work autonomously from the central command center which is hosted in the Cloud.

Next, we proposed the use of semantic agents. Agents are a natural choice to represent the drones, and adding the semantic components allows us to integrate and make analysis from data with any structure. Further, we noted that instead of having a single agent for each drone, we use mobile agents which also has the possibility to migrate from one drone to the other in case of need.

Then, because the data which is being processed has such a volume, velocity, and structure, it is not realistic to use standard tools. Hence, we end up with a Big Semantic Data problem. In order to solve this problem, we proposed the use of an

Evolving Knowledge Ecosystem which allows the integration of information from different sources in a manner similar to the workings of natural organisms.

The problem areas which were touched in this chapter were described from a high level perspective. There is still a very broad scope of topics which can be researched from both a theoretical and experimental perspective. Examples include, but are not limited to, what kind of robots are most useful, which factors influence the optimality of the system and up to which extend, how the communication between the robots and the Cloud should be designed, how the decision whether to perform a computation in the Cloud or on the devices locally should be taken, what type of agent system should be used, whether the evolving knowledge ecosystem has a performance high enough to accommodate the needs, and so on.

Acknowledgements The authors would like to thank the department of Mathematical Information Technology of the University of Jyväskylä for financially supporting this research. Further we would like to thank the members of the Industrial Ontologies Group (IOG) of the university of Jyväskylä for their support in the research.

References

Abadi DJ, Marcus A, Madden SR, Hollenbach K (2007) Scalable semantic web data management using vertical partitioning. In: Proceedings of the 33rd international conference on very large data bases. VLDB'07. VLDB Endowment, pp. 411–422

Akerkar R (ed) (2013) Big data computing. Taylor & Francis group-Chapman and Hall/CRC

Atmosukarto I, Ghanem B, Ahuja N (2012) Trajectory-based Fisher kernel representation for action recognition in videos. In: 2012 21st international conference on pattern recognition (ICPR), pp 3333–3336

Baader F, Calvanese D, McGuinness DL, Nardi D, Patel-Schneider PF (eds) (2003) The description logic handbook: theory, implementation, and applications. Cambridge University Press, Cambridge

Berners-Lee T, Hendler J, Lassila O et al (2001) The semantic web. Scientific american 284(5):28–37

Broekstra J, Kampman A, Harmelen Van F (2002) Sesame: a generic architecture for storing and querying rdf and rdf schema. In: The Semantic Web—ISWC 2002, pp 54–68

Buyya R, Yeo CS, Venugopal S, Broberg J, Brandic I (2009) Cloud computing and emerging IT platforms: Vision, hype, and reality for delivering computing as the 5th utility. Futur Gener Comput Syst 25(6):599–616

Cochez M (2012) Semantic agent programming language: use and formalization

Cochez M, Periaux J, Terziyan V, Kamlyk K, Tuovinen T (2014) Evolutionary cloud for cooperative UAV coordination. Software and Computational Engineering, No. C, Reports of the Department of Mathematical Information Technology, Series C, p 1

Damjanovic V, Kurz T, Westenthaler R, Behrendt W, Gruber A, Schaffert S (2011) Semantic enhancement: the key to massive and heterogeneous data pools. In: Proceeding of the 20th international IEEE ERK (electrotechnical and computer science) conference

Darwin C (1859) On the origins of species by means of natural selection. Murray, London

Dean D, Webb C (2011) Recovering from information overload. McKinsey Quarterly

Deelman E, Gannon D, Shields M, Taylor I (2009) Workflows and e-Science: an overview of workflow system features and capabilities. Futur Gener Comput Syst 25(5):528–540

Dias MB, Zlot R, Kalra N, Stentz A (2006) Market-based multirobot coordination: a survey and analysis. Proc IEEE 94(7):1257–1270

Doherty P, Heintz F, Kvarnström J (2013) High-level mission specification and planning for collaborative unmanned aircraft systems using delegation. Unmanned Syst 01(01):75–119

Ermolayev V, Akerkar R, Terziyan V, Cochez M (2013) Chap. Towards evolving knowledge ecosystems for big data understanding. In: Akerkar R (ed) Big data computing (2013)

Evangelinos C, Hill C (2008) Cloud Computing for parallel Scientific HPC applications: feasibility of running coupled atmosphere-ocean climate models on Amazon's EC2. Ratio, 2(2.40):2–34

Fayyad UM, Piatetsky-Shapiro G, Smyth P, et al (1996) Knowledge discovery and data mining: towards a unifying framework, vol. 96. KDD, pp 82–88

Ferber J (1999) Multi-agent systems: an introduction to distributed artificial intelligence, vol. 1. Addison-Wesley Reading

Fok K-L, Wong T-T, Wong M-L (2007) Evolutionary computing on consumer graphics hardware. IEEE Intell Syst 22(2):69–78

Gonzalez LF, Lee D-S, Walker RA, Periaux J (2009) Optimal mission path planning (MPP) for an air sampling unmanned aerial system. In: Scheding S (ed) 2009 Australasian conference on robotics & automation. Robotics & Automation Association, Sydney, Australian, pp 1–9

Goodfellow I, Bengio Y, Courville A (2016) Deep learning. MIT Press. Book in preparation, draft chapters. Available at http://www.deeplearningbook.org/

Haerder T, Reuter A (1983) Principles of transaction-oriented database recovery. ACM Comput Surv 15(4):287–317

Ito T, Chakraborty S, Kanamori R, Otsuka T (2012) Innovating multiagent algorithms for smart city: an overview. In: 2012 5th IEEE international conference on service-oriented computing and applications (SOCA), pp 1–8

Jackson KR, Ramakrishnan L, Muriki K, Canon S, Cholia S, Shalf J, Wasserman HJ, Wright NJ (2010) Performance analysis of high performance computing applications on the amazon web services cloud. In: 2010 IEEE second international conference on cloud computing technology and science (CloudCom). IEEE, pp 159–168

Kang K, Prasad JVR (2013) Development and flight test evaluations of an autonomous obstacle avoidance system for a rotary-wing UAV. Unmanned Syst 01(01):3–19

Katasonov A, Kaykova O, Khriyenko O, Nikitin S, Terziyan V (2008) Smart semantic middleware for the internet of things. In: Proceedings of the 5-th international conference on informatics in control, automation and robotics, pp 11–15

Kopeikin AN, Ponda SS, Johnson LB, How JP (2013) Dynamic mission planning for communication control in multiple unmanned aircraft teams. Unmanned Syst 01(01):41–58

Lassila O (1999) Resource description framework (RDF) model and syntax specification. W3C Recommendation. W3C. http://www.w3.org/TR/1999/REC-rdf-syntax-19990222/

LeCun Y, Bengio Y, Hinton G (2015) Deep learning. Nature 521(7553):436–444

Lee CA (2010) A perspective on scientific cloud computing. In: Proceedings of the 19th ACM international symposium on high performance distributed computing. ACM, pp 451–459

Li Z, Chen C, Wang K (2011) Cloud computing for agent-based urban transportation systems. IEEE Intell Syst 26(1):73–79

Matus R (2016) Neurala becomes largest supplier of deep learning software running in real time on a drone. (Sept 28)

Mell P, Grance T (2009) The NIST definition of cloud computing, version 15. Information Technology Laboratory. National Institute of Standards and Technology (NIST). http://www.csrc.nist.gov

Miller E, Manola F 2004 RDF primer. W3C Recommendation. W3C. http://www.w3.org/TR/2004/REC-rdf-primer-20040210/

Nardi D, Brachman RJ (2003) The description logic handbook: theory, implementation, and applications. In: Chap. An introduction to description logics, pp 1–40

Nash J (1951) Non-cooperative games. In: Annals of mathematics, pp 286–295

Nimmagadda Y, Kumar K, Lu Y-H, Lee CSG (2010) Real-time moving object recognition and tracking using computation offloading. In: 2010 IEEE/RSJ international conference on intelligent robots and systems (IROS), pp 2449–2455
Ostermann S, Iosup A, Yigitbasi N, Prodan R, Fahringer T, Epema D (2010) A performance analysis of EC2 cloud computing services for scientific computing. In: Cloud computing. Springer, pp 115–131
Pathak PH, Dutta R (2011) A survey of network design problems and joint design approaches in wireless mesh networks. IEEE Commun Surv Tutor 13(3):396–428
Peng D, Dabek F (2010) Large-scale incremental processing using distributed transactions and notifications. In: Proceedings of the 9th USENIX symposium on operating systems design and implementation
Petillot Y, Sotzing C, Patron P, Lane D, Cartright J (2009) Multiple system collaborative planning and sensing for autonomous platforms with shared and distributed situational awareness. In: Proceedings of the UAVSI's unmanned systems Europe
Sadashiv N, Kumar SMD (2011) Cluster, grid and cloud computing: a detailed comparison. In: 2011 6th international conference on computer science & Education (ICCSE). IEEE, pp 477–482
Simmhan Y, van Ingen C, Subramanian G, Li J (2010) Bridging the gap between desktop and the cloud for eScience applications. In: 2010 IEEE 3rd international conference on cloud computing (CLOUD). IEEE, pp 474–481
Song Z (2016) Cloud-based control system for unmanned aerial vehicles. US Patent App. 15/046,560. (Feb18)
Sorokin A, Forsyth D (2008) Utility data annotation with amazon mechanical turk. In: IEEE computer society conference on computer vision and pattern recognition workshops, 2008. CVPRW'08, pp 1–8
Sycara K, Norman TJ, Giampapa JA, Kollingbaum MJ, Burnett C, Masato D, McCallum M, Strub MH (2010) Agent support for policy-driven collaborative mission planning. Comput J 53(5):528–540
Taleb NN (2010) The black swan: the impact of the highly improbable fragility. Random House Digital, Inc
Terziyan V (2008) SmartResource-Proactive self-maintained resources in semantic web: Lessons learned. Int J Smart Home 2(2):33–57
Trigui S, Koubaa A, Ben Jamaa M, Chaari I, Al-Shalfan K (2012) Coordination in a multi-robot surveillance application using Wireless Sensor Networks. In: 2012 16th IEEE mediterranean electrotechnical conference (MELECON), pp 989–992
Vecchiola C, Pandey S, Buyya R (2009). High-performance cloud computing: a view of scientific applications. In: 2009 10th international symposium on pervasive systems, algorithms, and networks (ISPAN). IEEE, pp 4–16
Walker E (2008) Benchmarking Amazon EC2 for high-performance scientific computing. Usenix Login 33(5):18–23
Wang K, Shen Z (2011) Artificial societies and GPU-based cloud computing for intelligent transportation management. IEEE Intell Syst 26(4):22–28
Zhang Q, Cheng L, Boutaba R (2010) Cloud computing: state-of-the-art and research challenges. J Internet Serv Appl 1(1):7–18
Zikopoulos PC, Eaton C, deRoos D, Deutsch T, Lapis G (2012) Understanding big data. Mc Graw Hill

Part II
Computational Methods and Models

Chapter 4
A Simple Metaheuristic for the FleetSize and Mix Problem with TimeWindows

Olli Bräysy, Wout Dullaert and Pasi P. Porkka

Abstract This paper presents a powerful new single-parameter metaheuristic to solve the Fleet Size and Mix Vehicle Routing Problem with Time Windows. The key idea of the new metaheuristic is to perform a random number of random-sized jumps in random order through four well-known local search operators. Computational testing on the 600 large-scale benchmarks of Bräysy et al. (Expert Syst Appl 36(4):8460–8475, 2009) show that the new metaheuristic outperforms previous best approaches, finding 533 new best-known solutions. Despite the significant number of random components, it is demonstrated that the variance of the results is rather low. Moreover, the suggested metaheuristic is shown to scale almost linearly up to 1000 customers.

4.1 Introduction

The classical or capacitated Vehicle Routing Problem (VRP) assumes vehicles to be homogeneous and customers to accept service at any given time. However, these assumptions are rarely met in practice. On the other hand, considering heterogeneous vehicles is more practical and complex, but less studied in the literature. The growing interest in more complex variants of the VRP (Gendreau et al. 2008) has amongst others led to the formulation of different types of heterogeneous routing problems, which are classified and discussed in Baldacci et al. (2008).

O. Bräysy (✉)
Department of Mathematical Information Technology, University of Jyväskylä, P.O. Box 35, FI-40014 Jyväskylä, Finland
e-mail: olli.m.p.braysy@jyu.fi

W. Dullaert
Faculty of Economics and Business Administration, VU University Amsterdam, De Boelelaan 1105, 1081 HV Amsterdam, Netherlands
e-mail: wout.dullaert@vu.nl

P.P. Porkka
Justecon Oy, Fredrikinkatu 64 b 22, 00100 Helsinki, Finland
e-mail: pasi.p.porkka@gmail.com

P. Diez et al. (eds.), *Computational Methods and Models for Transport*, Computational Methods in Applied Sciences 45,
DOI 10.1007/978-3-319-54490-8_4

Liu et al. introduced time windows in heterogeneous routing problems and developed a benchmark of 168 instances derived from the well-known 100-customer Solomon benchmark for the VRPTW (Liu and Shen 1999; Solomon 1987). Their paper triggered a series of contributions to the so-called Fleet Size and Mix Vehicle Routing Problem with Time Windows (FSMVRPTW), as surveyed in Bräysy et al. (2008). In practice, vehicles often differ in their capacities, cost structures or equipment and considerable routing problems need to be solved daily. Academic research on heterogeneous routing problems has been limited to relatively small problem instances, often using the 100-customer problem instances of Liu and Shen (1999) as a benchmark. Bräysy et al. (2008) presented a deterministic annealing heuristic capable of obtaining 167 new best known solutions for the (Liu and Shen 1999) instances. In this paper, we focus on the new distance-based objective variant for the FSMVRPTW to replace the common en-route time objective introduced in Bräysy et al. (2008), and report on a powerful single parameter heuristic outperforming previous approaches on 600 new large-scale problem instances for the FSMVRPTW introduced in Bräysy et al. (2009). The suggested new metaheuristic is based on randomized jumps and search order that are used together with four well-known local search heuristics. The new method is described in more detail in the next section. The performance of the different heuristics on the large-scale problem instances for the FSMVRPTW is analyzed in Sect. 4.3. Finally, conclusions and directions for further research are formulated in Sect. 4.4.

4.2 Solution Approach

The suggested solution method consists of three phases and a preprocessing step. The preprocessing step is used to define a limiting value for each customer point, specifying a radius (distance) where the c closest customers are located. This speeds up the identification of the close customers during the search. In Phase 1, a variant of the well-known savings heuristic is used to generate the initial solutions. In Phase 2, the focus is on reducing the number of vehicles by means of a simple route elimination heuristic and in Phase 3, the obtained solution is further improved using four local search operators and the suggested Random Jump (RD) metaheuristic.

4.2.1 Phase 1: Initial Solution Construction

The single initial solution is generated with a modification of the savings heuristic (Clarke and Wright 1964). The search is started by serving each customer in a separate route. Then, an attempt is made to merge routes in a greedy way until the total cost of the solution (including both vehicle cost and total distance) cannot be improved. Vehicle sizes are updated whenever needed and always set to the smallest vehicle available capable of serving the customers on the route. When merging two routes

$R1$ and $R2$, the search is not limited to inserting $R1$ either in front of or after $R2$, but also considers positions between consecutive customers within route $R2$. To save computation time, the search is restricted so that only the c customers from $R2$ which are closest to the end points of $R1$ are considered as merging points.

4.2.2 Phase 2: Route Elimination Heuristic

In the second phase, the route elimination procedure (ELIM) is used to reduce the number of routes in the initial solution. ELIM considers all routes of the incumbent solution for depletion in random order, until no more improvements can be found with respect to the total cost objective.

For a given route, the current ELIM implementation removes all customers and tries to insert them one by one in the remaining routes according to their service sequence. For a given customer v and remaining route $R2$, only insertion positions adjacent to one of the c closest customers with regard to customer v are considered and the best feasible insertion according to the total cost objective is selected. In case the current route cannot be eliminated at a lower cost, the executed insertions are backtracked.

4.2.3 Phase 3: New Metaheuristic

4.2.3.1 Applied local searches

In Phase 3, an attempt is made to further improve the solution from Phase 2. The improvement is based on four local search heuristics. In addition to the aforementioned ELIM procedure, the applied local search operators include a route splitting operator called SPLIT as well as improved variants of the ICROSS and IOPT operators suggested in Bräysy (2003).

The SPLIT neighborhood consists of all solutions that result from splitting a single route in the current solution into two parts at any point. We employ it in a greedy, first-accept fashion, simply by looping through all routes and all customers in them, splitting the selected route into two parts at the position of the current customer. The move is made if the split reduces the total cost. Here the use of SPLIT and ELIM is restricted so that they are not used at the same iteration to enable better search diversification.

ICROSS relocates or exchanges the segments of consecutive customers between two separate routes, such that it also inverts the customers' order in considered segments. The ICROSS neighborhood is usually large, so two limitation strategies are applied to achieve scalability. The first limitation is the maximum segment length ls that is determined randomly within the 1–10 range. The second limitation is that the segment insertion position in another route must always be among the c closest customers. The IOPT intra-tour operator is a generalization of Or-opt (Or 1976). As ICROSS, it also includes moves where the segment is reversed and it applies the same two limitation strategies. Here both ICROSS and IOPT are modified so that they consider the different vehicle properties and costs and possible adjustment of the vehicle type as a part of the move.

4.2.3.2 Metaheuristic framework

The main framework of the suggested new metaheuristic reminds well-known Simulated Annealing (SA) (Kirkpatrick et al. 1983). The key idea is to allow also local search moves, which may worsen the objective function value and alternate between search phases where worsening moves are allowed and where they are not allowed. Here, instead of setting a fixed limit for the maximum worsening of an individual move and a scheme for adjusting the limit, we define a random limit S for the maximum worsening of an individual move at the beginning of each iteration. More precisely, the value of S is determined randomly in the 0.1–50% range, referring to the allowed increase of the single route objective value. In addition to setting the limits on the deterioration of a single move, we also define a limit M on the maximum number of deteriorating moves. After the maximum number of deteriorating moves have been executed, only improving moves are allowed until improvements can be found. The value of M is determined randomly in the range 1–100 each time the search reaches a local minimum. To further diversify the search, the routes of the current solution and the points in the selected routes are considered for improvement in random order and the first feasible move that fits the acceptance criteria is implemented in each step. For the SPLIT neighborhood, we did not allow moves that could worsen objective value, as we found that it resulted in too large diversifications.

If no more improvements have been found for a given number of iterations I, the search is restarted from the best solution found by far. Here the value of I is chosen randomly in the range 1–10% of the total number of iterations N.

4.3 Computational Experiments

4.3.1 Problem Data and Experimental Setting

In contrast to Liu and Shen (1999), who minimized the sum of all vehicle costs and en-route time, we consider the sum of all the vehicle costs and total distance as optimization objective. The new objective was first introduced in Bräysy et al. (2008) and it has been believed to be of a higher practical value than the former objective function. To support the research on large-scale heterogeneous routing problems, (Bräysy et al. 2009) developed a new set of benchmark problems based on the large-scale VRPTW benchmark instances of Gehring and Homberger (1999).

Table 4.1 Cost structure A

C1		C2		R1	
Cost	Capacity	Cost	Capacity	Cost	Capacity
40	200	120	575	40	140
70	335	240	1100	70	230
100	460	350	1540	100	310
140	615	470	1975	140	405
170	715	580	2320	170	460
200	800	700	2700	200	500
240	910	820	2955	240	550
270	975	930	3160	270	565
R2		RC1		RC2	
170	590	40	125	170	590
340	1115	70	205	340	1115
500	1550	100	275	500	1550
670	1945	140	355	670	1945
840	2270	170	420	840	2270
1000	2500	200	450	1000	2500
1170	2690	240	495	1170	2690
1330	2795	270	500	1330	2795

For 300 benchmark problems, ranging from 200 to 1000 customers each, a single set of 8 vehicle types and 2 cost levels were constructed based on real-life cost data and general testing considerations.

Table 4.1 summarises the cost structure A of the extended benchmarks. As in Liu and Shen (1999), we calculate cost structure C by dividing the vehicle costs of cost structure A by 10, resulting in 600 test instances with both relatively low and relatively high vehicle costs.

We use here the same general search limitation strategies and implementation details as in Bräysy et al. (2008, 2009), i.e., the double-linked list in an array data structure (Kytöjoki et al. 2007), standard time window feasibility check techniques (Campbell and Savelsbergh 2004), and the opportunistic constraint feasibility check strategy. The latter means that features that are the easiest or quickest to calculate or that are the most likely to violate a feasibility condition, are checked first. The applied algorithms were programmed with Java and tested on Intel Core I7 (2.93 GHz) computer with 6 GB memory.

4.3.2 *Sensitivity Analysis*

In this section we illustrate the impact of the closeness limit c on the solution quality and the importance of the key components of the suggested method. Both analyses are based on running the algorithm on the entire problem set with each value and fixing the random components. The first issue is demonstrated in Fig. 4.1. Based on Fig. 4.1, it is rather surpricing how small an impact the different values have and that values above 15 do not improve the solution quality. Obviously, high quality solutions can be obtained even if the search is strictly restricted to the closest positions only. For the actual test runs, value 15 was chosen.

Table 4.2 demonstrates the importance of the key components of the suggested algorithm. Again, the values are based on executing the algorithm on the entire problem set. The variance refers to the average percentage difference between the best and the worst value obtained after running the algorithm 10 times on each instance. The second column illustrates the average difference in the solution quality if the routes are not put into random order in the beginning of each iteration. Correspondingly, the third column shows the solution quality impact of considering the points of selected routes in random order for improvement. The fourth column depicts the value of having the restarts from the best-found solution and the last column the importance of the splitting operator.

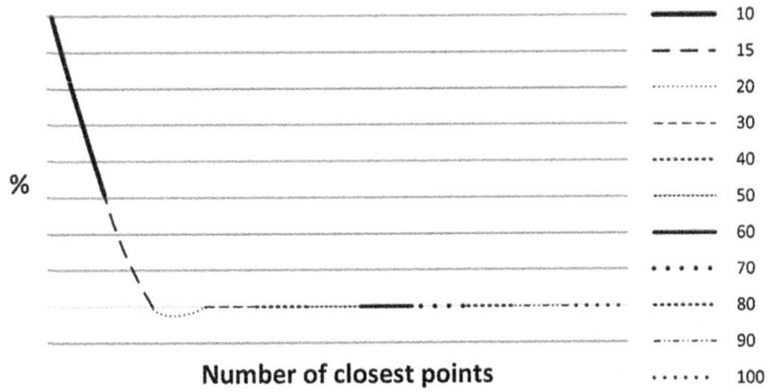

Fig. 4.1 The impact of closeness limit to solution quality

Table 4.2 Average impact of key algorithm components

Variance	No route ordering	No move ordering	No restarts	No splits
0.74%	0.35%	1.51%	1.06%	0.18%

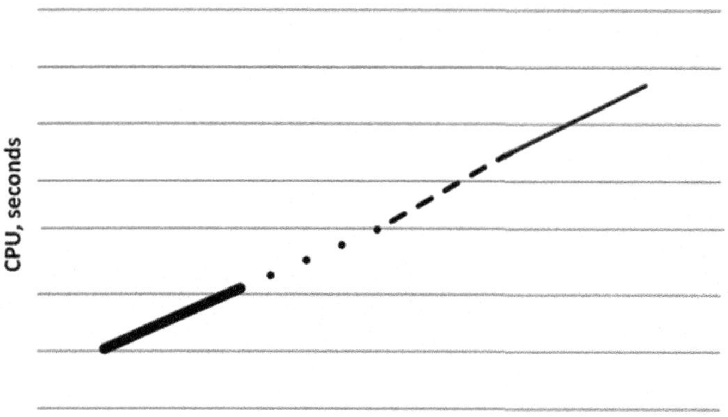

Fig. 4.2 The scaling of the method

Based on Table 4.2, the variance of the results is surprisingly low, considering the significant amount of random components, indicating the efficiency and robustness of the solution method. It also appears that putting moves to random order and considering the restarts are important to obtain high-quality solutions. As the random-sized jumps and CROSS and IOPT operators are at the core of the new method, their role is obviously crucial. Thus, their impact was not separately tested in the sensitivity analysis. The scaling of the CPU times over the tested problem sizes is illustrated in Fig. 4.2, which shows that the scaling is practically linear.

4.3.3 Comparison of Results

Computational testing on both benchmark sets is summarized in Tables 4.3 and 4.4 against the best two competing and published methods by Bräysy et al. (2008, 2009), respectively. Here the CPU times are normalized so that they take into account the number of executed runs to obtain the reported solution and the differences in

Table 4.3 Results to cost structure A

Problem size	BDHG			BPDRT			NEW		
	Objective	Gap	CPU	Objective	Gap	CPU	Objective	Gap	CPU
200	13494.78	0.51	51.99	13588.70	1.21	33.55	13232.84	−1.44	10.62
400	28121.16	0.63	220.03	28337.82	1.41	64.70	27555.87	−1.38	20.97
600	46648.79	1.15	481.46	46498.11	0.83	102.82	45365.17	−1.63	31.57
800	67325.11	1.85	854.72	66312.86	0.32	139.83	65495.94	−0.91	45.35
1000	91862.62	3.11	1267.79	89165.95	0.09	191.26	88399.03	−0.77	57.72

Table 4.4 Results to cost structure C

Problem size	BDHG			BPDRT			NEW		
	Objective	Gap	CPU	Objective	Gap	CPU	Objective	Gap	CPU
200	3870.58	1.21	56.32	3829.72	0.14	36.80	3763.18	−1.60	11.12
400	8658.01	1.93	231.04	8502.17	0.13	68.69	8306.29	−2.19	21.29
600	16592.86	3.51	486.91	16036.31	0.11	94.63	15631.60	−2.49	32.02
800	26940.42	3.74	901.49	25994.01	0.09	141.09	25349.44	−2.39	45.27
1000	40557.01	5.28	1439.63	38552.68	0.08	189.12	37407.77	−2.90	60.55

computer speeds. The latter is done simply by running all algorithms directly on the same computer. The gap refers to average percentage difference compared to the best-known solutions.

The results clearly show that the new method provides better solutions in all problem sizes with less computational effort. The differences appear to be larger with cost structure C where vehicle costs have smaller impact. The previous best-known solutions are on average improved by more than 2% and the largest differences in solution quality between the compared methods are even above 8%. Overall, the suggested new method found 535 new best-known solutions to 600 benchmark instances. The detailed results are provided in Appendix I. When interpreting the results, one must also note that the considered two competing methods have clearly outperformed the other FSMVRPTW heuristics on the Liu and Shen problem set.

4.4 Conclusions

Large scale heterogeneous routing problems are common in practice and have only recently received academic attention. In this paper, we have suggested a new powerful metaheuristic for the Fleet Size and Mix Routing Problem with Time Windows. We have shown that simple strategy of defining the size and number of deteriorating moves randomly outperforms traditional metaheuristic strategies. The suggested method is also very fast, scales linearly with problem size and have found 535 new best-known solutions to benchmark problems.

Appendix

Data set	MSDAL		Normal		New		Data set	MSDAL		Normal		New	
	Best	CPU	Best	CPU	Best	CPU		Best	CPU	Best	CPU	Best	CPU
C121A	17478.35	95.30	17456.15	7.55	**16146.77**	14.32	C121C	4216.08	106.09	4216.08	8.34	**3801.05**	19.15
C122A	17056.41	96.25	17095.62	7.58	**16929.05**	10.69	C122C	4165.48	119.16	4164.70	7.58	**4162.99**	13.53
C123A	16897.70	92.95	16820.15	7.86	**16629.48**	14.45	C123C	4104.72	114.55	4093.93	8.72	**4068.95**	14.80
C124A	16563.86	95.13	16529.97	6.58	**16364.38**	16.10	C124C	4009.70	108.95	3988.11	7.89	**3966.33**	15.27
C125A	17452.85	108.22	17450.12	6.91	**17311.40**	8.07	C125C	4210.44	114.11	**4210.35**	7.58	**4210.35**	8.85
C126A	17528.25	114.49	17466.85	8.09	**17407.42**	10.83	C126C	**4210.35**	108.91	**4210.35**	8.47	**4210.35**	10.17
C127A	17539.70	106.81	17427.81	7.80	**17280.28**	10.31	C127C	4196.00	125.78	**4191.71**	7.84	**4191.71**	9.69
C128A	17366.06	107.13	17284.68	9.03	**17118.19**	11.56	C128C	4165.85	121.14	4169.79	7.84	**4165.50**	12.64
C129A	16963.26	85.86	17018.38	7.64	**16810.15**	12.17	C129C	4106.18	131.17	4086.01	7.31	**4069.22**	10.39
C1210A	16884.73	102.50	16749.52	7.41	**16618.35**	12.90	C1210C	4051.22	102.95	4038.56	8.11	**4001.34**	13.67
C221A	17787.35	69.27	17901.59	6.53	**17281.46**	8.10	C221C	**3530.70**	82.17	3547.22	7.75	3533.53	9.31
C222A	17591.14	54.52	17649.04	6.09	**16955.05**	9.81	C222C	3465.22	70.08	3462.97	7.39	**3420.01**	8.30
C223A	17139.14	62.72	16984.26	6.28	**16946.04**	12.50	C223C	3397.72	75.94	3396.56	7.05	**3356.29**	11.69
C224A	17048.28	86.38	17177.17	5.83	**16823.39**	9.21	C224C	3381.71	89.45	3333.64	6.83	**3294.95**	10.81
C225A	17627.05	58.06	17567.03	6.14	**17175.93**	9.34	C225C	3483.03	67.34	3496.70	7.14	**3462.77**	7.74
C226A	17321.54	65.31	17415.59	6.25	**16999.21**	6.19	C226C	3447.47	74.23	3439.13	7.39	**3415.59**	7.30
C227A	17496.18	66.75	17477.99	6.38	**17015.52**	8.08	C227C	3507.58	67.89	3491.72	7.11	**3419.37**	10.05
C228A	17399.34	60.69	17177.48	7.31	**16890.76**	8.96	C228C	3451.61	69.92	3425.01	7.58	**3372.27**	7.27
C229A	17600.02	62.81	17478.95	5.98	**16991.30**	9.88	C229C	3481.13	78.67	3439.75	7.06	**3415.39**	9.39
C2210A	17324.16	63.83	17024.78	6.19	**16805.04**	10.28	C2210C	3363.75	68.09	3408.34	6.92	**3334.50**	6.49
R121A	13904.64	90.36	13897.19	5.78	**13790.32**	9.19	R121C	5713.38	100.48	5700.76	6.03	**5669.65**	9.67
R122A	12467.81	84.94	12367.64	7.36	**12234.33**	11.59	R122C	4931.09	85.88	4889.25	7.59	**4833.51**	14.40
R123A	11207.92	80.00	11122.32	6.80	**11026.50**	12.98	R123C	4211.11	85.06	4125.01	7.48	**4092.97**	13.26
R124A	10466.36	74.38	10388.10	6.28	**10267.04**	13.00	R124C	3576.54	93.22	3517.42	7.41	**3497.69**	12.92
R125A	12336.34	81.27	12294.94	7.33	**12110.40**	7.60	R125C	4974.72	89.55	4941.12	7.69	**4857.98**	10.50
R126A	11620.63	79.56	11332.67	6.74	**11269.22**	10.67	R126C	4331.41	80.02	4340.97	6.94	**4283.74**	10.92
R127A	10868.82	76.19	10669.74	7.23	**10569.21**	9.89	R127C	3823.36	80.72	3759.99	8.06	**3719.42**	12.09
R128A	10361.97	72.53	10216.89	6.22	**10167.08**	12.18	R128C	3445.34	81.13	3421.79	7.70	**3362.84**	14.88
R129A	11812.63	73.99	11639.96	7.53	**11443.92**	8.96	R129C	4567.00	77.06	4553.55	7.69	**4471.58**	8.96
R1210A	10890.60	71.27	10780.94	7.39	**10641.88**	10.20	R1210C	3922.18	75.00	3867.58	7.25	**3817.39**	11.61
R221A	**14355.06**	71.61	14840.05	6.30	14874.75	6.01	R221C	4972.08	52.34	4805.18	6.98	**4758.35**	12.62
R222A	**13543.70**	52.16	14209.18	6.19	13799.69	12.36	R222C	4526.72	63.00	4344.74	6.64	**4258.26**	13.06
R223A	12103.17	57.30	12706.39	6.28	**12008.67**	10.52	R223C	3984.50	61.58	3795.58	6.97	**3767.15**	13.45
R224A	10520.72	57.80	11122.21	6.11	**10514.71**	11.39	R224C	3143.03	72.02	3097.62	6.98	**2887.55**	10.37
R225A	**13081.80**	38.53	13444.59	6.48	13383.36	10.97	R225C	4369.26	54.05	4375.84	6.77	**4278.22**	7.49
R226A	12543.67	48.02	12599.55	6.20	**12098.39**	10.97	R226C	4046.82	49.58	3947.06	7.05	**3873.87**	11.86
R227A	**11137.18**	52.20	11657.07	6.33	11325.15	11.65	R227C	3648.21	58.73	3477.60	6.69	**3409.74**	11.67
R228A	10274.93	65.78	10717.99	6.59	**10004.11**	11.59	R228C	2935.01	73.13	2896.26	7.23	**2792.17**	9.83
R229A	12586.96	34.84	12636.17	6.24	**12183.42**	9.39	R229C	4173.26	44.49	4138.39	7.36	**4026.33**	11.54
R2210A	11347.19	41.95	12182.76	6.08	**11263.86**	10.25	R2210C	3843.46	36.20	3817.64	6.42	**3672.94**	10.34
RC121A	10995.16	94.88	10885.35	6.52	**10814.30**	7.15	RC121C	4239.52	95.44	4239.69	7.24	**4181.93**	10.20
RC122A	10413.91	83.36	10214.64	6.94	**10103.06**	8.75	RC122C	3919.43	91.63	3882.92	7.33	**3828.64**	9.55
RC123A	9909.46	77.89	9644.62	6.97	**9590.68**	12.43	RC123C	3587.59	81.97	3498.29	7.53	**3431.44**	11.09
RC124A	9494.38	74.58	9378.79	6.28	**9260.15**	13.96	RC124C	3289.12	93.98	3197.80	7.81	**3167.56**	13.42
RC125A	10603.03	93.73	10509.68	7.41	**10299.01**	10.33	RC125C	4021.30	89.45	3984.16	7.70	**3932.96**	9.75
RC126A	10526.77	86.44	10368.06	7.47	**10145.36**	8.42	RC126C	3945.98	96.44	3926.56	8.06	**3839.79**	11.51
RC127A	10179.64	83.75	10030.10	7.83	**9941.54**	10.17	RC127C	3799.75	88.03	3728.42	7.25	**3676.90**	9.52
RC128A	9856.77	77.39	9714.63	7.42	**9655.27**	10.19	RC128C	3552.67	85.84	3564.54	7.78	**3472.88**	12.73
RC129A	9813.89	80.84	9737.70	7.56	**9609.07**	10.28	RC129C	3532.69	85.20	3541.33	7.47	**3452.41**	12.01
RC1210A	9645.79	74.55	9558.28	7.49	**9425.92**	9.77	RC1210C	3358.61	76.81	3361.24	7.61	**3306.65**	11.64
RC221A	14529.34	61.31	14520.71	6.09	**14427.00**	4.74	RC221C	4149.91	62.95	4114.66	7.09	**4047.60**	6.05
RC222A	13425.73	64.33	13705.36	5.89	**13105.64**	14.07	RC222C	3807.73	67.23	3784.42	7.49	**3687.16**	9.74
RC223A	11976.30	64.14	12827.49	5.53	**11738.56**	7.15	RC223C	3701.30	78.53	3498.41	7.03	**3386.92**	8.14
RC224A	**10811.14**	91.09	11279.77	6.16	10850.91	12.59	RC224C	3219.43	101.28	2980.75	6.66	**2968.82**	13.02
RC225A	13392.72	51.97	14359.90	5.52	**13170.90**	11.47	RC225C	3887.52	51.31	3769.78	6.91	**3684.16**	13.00
RC226A	13295.09	45.20	13565.32	5.63	**12916.40**	13.31	RC226C	3820.37	52.39	3761.27	7.31	**3663.62**	13.35
RC227A	**12345.43**	52.80	12879.30	6.33	12363.69	11.36	RC227C	3545.04	57.17	3572.29	7.41	**3463.24**	12.07
RC228A	11989.91	55.75	12449.37	5.33	**11600.87**	12.25	RC228C	3440.13	54.98	3351.74	7.31	**3244.29**	12.22
RC229A	11640.05	41.98	12109.84	8.09	**11136.85**	14.42	RC229C	3326.45	63.41	3257.88	6.84	**3165.38**	11.03
RC2210A	11345.16	51.19	11603.96	5.30	**10339.85**	9.25	RC2210C	3005.69	48.02	3142.97	6.97	**2984.82**	9.02

(continued)

(continued)

Data set	MSDAL		Normal		New		Data set	MSDAL		Normal		New	
	Best	CPU	Best	CPU	Best	CPU		Best	CPU	Best	CPU	Best	CPU
C141A	37156.67	368.98	37097.84	13.89	**37004.00**	10.30	C141C	10186.68	366.30	10185.25	14.16	**10183.32**	12.71
C142A	36287.65	375.16	36456.97	15.70	**36095.64**	18.77	C142C	10084.39	382.86	10130.08	14.64	**10067.15**	22.07
C143A	35443.86	366.56	35281.26	14.61	**34936.70**	23.37	C143C	9703.27	370.94	9848.58	15.34	**9683.62**	29.11
C144A	34483.09	353.73	33896.21	15.25	**33646.05**	28.74	C144C	9149.61	366.09	9134.87	15.75	**8999.95**	29.38
C145A	37423.25	375.06	37323.54	14.47	**36985.55**	17.05	C145C	10179.76	392.80	**10178.79**	13.89	**10178.79**	17.92
C146A	37640.19	389.23	37632.95	16.36	**37006.93**	17.80	C146C	**10177.34**	384.77	10181.25	14.78	**10177.34**	20.90
C147A	37628.29	368.61	37656.93	16.64	**37157.52**	18.86	C147C	10162.34	432.42	10260.60	16.11	**10162.03**	19.63
C148A	37268.20	356.30	37174.88	18.25	**36643.17**	20.09	C148C	10121.30	407.55	10150.06	17.09	**10078.15**	23.57
C149A	36408.62	371.00	36532.43	16.80	**35952.34**	25.65	C149C	9949.86	405.22	9968.79	16.86	**9832.26**	26.16
C1410A	35691.67	354.22	35847.06	17.11	**35442.90**	27.16	C1410C	9660.06	381.63	9763.37	17.48	**9580.12**	27.00
C241A	37214.21	305.72	37088.94	9.75	**35018.11**	16.01	C241C	7574.53	339.64	7511.11	12.08	**7265.86**	16.86
C242A	35332.57	296.28	34896.10	11.27	**34394.99**	18.63	C242C	7507.78	324.50	7289.22	13.11	**7087.05**	18.41
C243A	35391.65	296.09	35503.71	10.48	**34322.87**	20.97	C243C	7460.24	337.22	7092.95	12.33	**7020.11**	19.86
C244A	34815.47	358.92	34546.07	10.81	**33721.40**	20.08	C244C	7172.85	387.73	6943.14	12.81	**6816.11**	24.32
C245A	37085.14	297.52	36290.36	10.78	**34785.97**	17.38	C245C	7493.45	317.08	7367.54	12.70	**7152.30**	16.68
C246A	36175.26	291.80	35696.75	11.72	**34489.11**	19.05	C246C	7331.15	324.31	7182.54	13.78	**7036.19**	18.49
C247A	36025.56	304.25	36003.90	11.55	**34384.41**	19.00	C247C	7407.89	316.39	7387.50	12.80	**7140.91**	17.33
C248A	35900.06	307.61	35860.75	10.55	**34167.01**	17.89	C248C	7317.89	306.81	7243.95	12.75	**6922.05**	16.83
C249A	36117.04	320.34	36265.80	10.64	**34482.65**	20.52	C249C	7395.61	325.31	7192.29	13.11	**7072.88**	21.11
C2410A	35373.19	321.88	35217.67	10.52	**33999.22**	16.99	C2410C	7076.80	327.58	7156.69	12.47	**6826.55**	18.24
R141A	28667.92	329.58	28416.94	10.45	**28106.24**	17.72	R141C	12254.98	370.05	12212.22	11.89	**12152.43**	19.69
R142A	25397.84	318.11	25291.12	13.14	**24902.38**	21.37	R142C	10502.18	349.16	10560.12	14.31	**10332.34**	25.94
R143A	23134.81	297.03	22818.20	12.55	**22602.52**	26.43	R143C	8956.00	318.00	8795.12	14.61	**8672.50**	28.85
R144A	21970.88	303.78	21715.80	12.41	**21503.57**	26.97	R144C	8043.02	343.14	7870.61	14.59	**7747.60**	30.97
R145A	25347.71	317.28	25250.04	13.11	**24774.88**	16.47	R145C	10608.92	334.73	10656.56	14.33	**10487.93**	18.85
R146A	23905.30	301.91	23641.90	15.19	**23373.36**	19.55	R146C	9713.02	327.27	9633.82	15.55	**9475.35**	23.32
R147A	22687.50	301.45	22352.49	13.59	**22121.96**	25.02	R147C	8555.58	323.72	8417.54	13.22	**8236.43**	25.43
R148A	21909.69	298.95	21606.92	13.14	**21444.98**	29.13	R148C	7905.12	335.55	7772.45	14.66	**7630.77**	25.96
R149A	24180.08	297.16	24065.69	13.56	**23661.41**	17.94	R149C	9921.21	318.06	9832.68	14.64	**9670.65**	20.75
R1410A	23130.23	285.44	22884.25	13.03	**22603.72**	21.92	R1410C	9078.38	308.25	8944.72	12.83	**8793.32**	23.20
R241A	**30752.74**	321.22	31463.53	13.16	31502.10	22.75	R241C	10799.49	185.02	10370.89	13.03	**10066.55**	19.47
R242A	**27372.86**	220.03	30025.97	12.03	28893.61	28.46	R242C	9635.11	267.55	9171.98	12.80	**9010.48**	21.33
R243A	**24922.94**	264.44	27621.15	11.81	25112.82	15.38	R243C	8338.94	330.72	8089.29	11.75	**7779.07**	14.93
R244A	**21552.34**	331.88	21894.05	10.66	21645.36	19.66	R244C	6809.50	306.08	6493.74	13.39	**6283.62**	23.13
R245A	**27302.03**	179.00	30007.57	11.42	28472.48	18.74	R245C	9644.09	204.39	9447.63	12.55	**9028.41**	17.61
R246A	**24400.15**	183.64	26785.61	11.17	25635.31	21.90	R246C	9092.40	240.33	8528.13	12.92	**8189.75**	17.41
R247A	**22749.95**	248.61	24206.06	10.49	22908.83	22.82	R247C	8004.41	269.05	7449.61	12.17	**7137.50**	17.55
R248A	21298.98	362.19	20392.97	10.58	**20334.21**	16.19	R248C	6552.64	313.17	6176.44	14.64	**5950.25**	15.49
R249A	**25891.73**	155.73	27132.32	11.38	26350.33	23.49	R249C	8991.19	196.50	8864.43	12.95	**8435.10**	24.06
R2410A	**23894.38**	184.95	25154.26	12.39	24447.99	26.19	R2410C	8567.66	200.52	8386.89	13.20	**7923.03**	19.13
RC141A	23667.18	348.45	23429.93	12.98	**23161.68**	12.79	RC141C	9839.00	360.55	9812.02	14.42	**9708.76**	17.25
RC142A	22066.80	327.91	21860.59	13.50	**21533.88**	18.77	RC142C	8970.15	349.41	8859.79	15.70	**8699.61**	21.87
RC143A	20597.40	316.80	20446.10	12.78	**20142.14**	24.04	RC143C	8140.88	337.50	8089.51	15.17	**7924.20**	25.69
RC144A	20080.59	331.70	19809.80	13.38	**19672.56**	26.75	RC144C	7723.72	361.88	7558.39	14.28	**7470.42**	28.39
RC145A	22271.61	341.41	22197.31	14.45	**21979.77**	17.38	RC145C	9306.67	341.63	9147.86	15.50	**8985.27**	19.38
RC146A	22257.21	338.55	21760.79	14.47	**21678.32**	19.77	RC146C	9150.17	338.03	9127.43	15.36	**8904.58**	16.77
RC147A	21663.84	327.77	21460.28	15.36	**21218.09**	20.44	RC147C	8844.62	328.13	8754.82	13.91	**8557.48**	20.70
RC148A	21004.48	313.55	20879.27	14.20	**20577.53**	21.26	RC148C	8391.29	323.69	8427.69	14.56	**8197.76**	22.64
RC149A	20841.74	311.05	20742.67	15.06	**20401.44**	21.61	RC149C	8297.20	309.86	8318.95	13.50	**8099.14**	23.76
RC1410A	20439.53	306.94	20197.91	12.97	**20101.92**	23.35	RC1410C	8085.39	320.97	8025.24	13.89	**7838.42**	23.60
RC241A	30354.46	348.42	29970.85	12.50	**29280.00**	23.96	RC241C	9035.96	267.58	8849.39	11.27	**8558.49**	22.83
RC242A	**26847.73**	341.09	28733.62	11.44	26869.42	20.62	RC242C	8567.81	356.14	7984.16	11.16	**7856.42**	22.51
RC243A	24497.60	345.91	26844.11	10.47	**23597.58**	22.67	RC243C	7763.58	361.31	7137.48	12.00	**6910.54**	23.10
RC244A	21047.48	359.88	**20679.44**	14.63	20859.13	19.13	RC244C	6609.23	375.17	5769.86	13.31	**5621.26**	21.03
RC245A	28217.45	264.09	28456.96	11.42	**27225.33**	22.84	RC245C	8289.54	272.08	8142.32	12.47	**7793.09**	21.26
RC246A	27157.56	292.84	28627.03	11.56	**26486.60**	21.81	RC246C	8063.89	302.53	8022.63	11.59	**7695.75**	20.76
RC247A	26655.32	247.58	27342.13	11.20	**25504.91**	23.28	RC247C	7724.78	259.39	7589.56	12.14	**7306.26**	21.14
RC248A	25030.18	250.05	25521.11	12.03	**23869.02**	22.81	RC248C	7338.61	257.86	7197.08	13.39	**6922.08**	21.70
RC249A	23893.36	255.91	23348.97	13.08	**22760.50**	21.68	RC249C	7235.74	287.03	7003.77	13.14	**6664.64**	19.20
RC2410A	23346.33	231.13	22963.14	16.47	**21397.63**	20.86	RC2410C	6931.63	258.11	6567.66	13.45	**6377.35**	14.01

(continued)

(continued)

Data set	MSDAL		Normal		New		Data set	MSDAL		Normal		New	
	Best	CPU	Best	CPU	Best	CPU		Best	CPU	Best	CPU	Best	CPU
C161A	59689.55	760.17	59586.47	21.09	**59481.40**	18.72	C161C	18707.09	793.52	18716.49	19.03	**18699.39**	21.22
C162A	58755.09	785.00	58571.76	24.34	**58079.81**	35.99	C162C	18526.04	793.52	18502.45	22.34	**18454.09**	37.94
C163A	57563.30	771.61	56829.20	24.06	**56346.77**	43.54	C163C	17911.72	799.81	17970.54	22.67	**17700.27**	48.14
C164A	56426.93	783.73	55567.63	22.53	**55186.06**	48.70	C164C	17264.42	779.80	17123.40	22.58	**16981.84**	48.02
C165A	59855.49	757.41	59943.50	21.64	**59422.74**	23.34	C165C	18706.06	822.45	18693.84	17.95	**18686.38**	26.47
C166A	60653.99	801.13	60318.27	23.66	**59381.65**	30.25	C166C	18717.27	813.41	18707.56	21.69	**18686.50**	31.57
C167A	60286.25	776.58	60294.54	24.36	**59443.34**	29.58	C167C	18704.18	863.47	18701.59	21.66	**18683.86**	32.06
C168A	59920.91	776.97	59595.55	25.70	**58996.74**	36.13	C168C	18571.11	801.09	18682.18	23.31	**18510.70**	36.61
C169A	58156.55	786.81	58567.22	27.03	**57804.33**	38.11	C169C	17887.99	815.83	18242.30	23.27	**17763.93**	41.73
C1610A	57870.92	754.78	57896.52	27.27	**57194.78**	43.93	C1610C	**17639.89**	815.38	17887.74	24.09	17646.48	45.05
C261A	58825.82	714.53	58082.40	17.44	**55132.04**	26.85	C261C	13015.11	716.86	13332.62	15.05	**12604.76**	23.92
C262A	58081.86	741.48	57698.77	19.27	**54475.36**	26.47	C262C	13020.56	742.78	12918.37	18.03	**12457.44**	28.70
C263A	55943.84	775.75	55292.05	16.86	**53935.22**	29.47	C263C	12849.37	779.22	12715.31	16.64	**12328.78**	31.97
C264A	55141.06	863.83	54421.80	16.83	**52625.95**	33.85	C264C	12532.26	830.28	12377.14	16.34	**11918.71**	35.80
C265A	58023.94	714.28	57227.76	19.17	**54984.43**	24.76	C265C	13424.80	703.59	13080.26	17.11	**12461.79**	24.76
C266A	57196.13	738.19	56603.26	17.83	**54784.32**	26.93	C266C	13769.02	745.63	13050.43	17.91	**12337.14**	22.76
C267A	57811.18	739.98	56829.07	19.36	**54898.39**	29.03	C267C	13399.36	711.77	13223.82	17.24	**12577.49**	24.41
C268A	56595.38	776.39	56119.74	18.13	**54172.38**	27.21	C268C	13038.78	764.05	12680.30	17.27	**12237.35**	26.69
C269A	56210.82	761.52	56183.81	19.19	**54197.74**	27.92	C269C	13054.19	754.39	12995.09	17.88	**12148.71**	26.55
C2610A	55862.66	735.16	55607.16	18.30	**53561.63**	26.63	C2610C	13189.45	753.11	12730.64	16.95	**12113.26**	30.75
R161A	48768.16	724.11	48035.25	16.31	**47810.70**	28.94	R161C	24292.87	751.25	24277.93	15.97	**23991.86**	35.52
R162A	43036.65	683.81	42172.33	21.94	**42115.37**	36.02	R162C	20600.75	697.09	20575.01	20.86	**20243.80**	38.00
R163A	39796.18	668.28	39089.63	21.48	**38733.18**	40.65	R163C	18382.28	691.66	17908.61	20.52	**17597.84**	43.62
R164A	37493.03	683.09	36769.21	20.94	**36402.21**	49.48	R164C	16321.85	718.95	15765.01	19.45	**15657.10**	45.60
R165A	43303.32	671.16	42517.05	23.97	**41881.58**	27.78	R165C	21421.98	674.55	21118.63	18.58	**20607.54**	26.61
R166A	41316.95	658.94	40314.96	21.25	**39953.39**	37.32	R166C	19323.65	672.50	19015.45	20.27	**18759.35**	33.51
R167A	38989.63	672.41	38234.20	20.67	**37851.14**	41.22	R167C	17921.75	681.05	17025.78	20.81	**17023.24**	40.89
R168A	36982.48	694.75	36390.44	21.97	**36175.07**	45.49	R168C	16051.66	722.38	15616.18	21.83	**15418.46**	41.71
R169A	41915.91	654.11	40960.30	23.81	**40522.86**	29.89	R169C	20085.27	656.09	19767.32	20.03	**19380.68**	30.42
R1610A	40068.23	637.75	39521.70	23.38	**39093.56**	30.67	R1610C	18891.04	652.08	18449.85	19.89	**18111.24**	36.10
R261A	55926.79	826.92	53192.13	20.61	**51291.20**	28.98	R261C	20981.80	470.61	19697.79	17.81	**19035.79**	29.30
R262A	48401.15	473.86	49406.02	18.70	**46659.77**	36.07	R262C	18525.78	562.38	17182.59	17.03	**16909.31**	32.01
R263A	40919.19	586.42	44327.34	17.02	**40101.74**	31.67	R263C	16550.00	616.45	14510.42	17.45	**14184.89**	33.37
R264A	35706.92	678.59	35083.09	17.39	**33980.67**	26.96	R264C	13157.25	707.73	11449.41	17.55	**11008.94**	27.53
R265A	**49199.77**	372.17	50693.53	20.89	49223.50	31.12	R265C	19533.05	447.58	18294.03	17.56	**17771.90**	33.01
R266A	**42134.19**	405.91	46572.58	17.70	43295.07	31.87	R266C	18749.86	404.58	16356.36	17.83	**15661.50**	30.06
R267A	39024.05	526.75	42750.32	16.58	**38492.23**	23.85	R267C	14898.00	544.27	13805.47	15.74	**13255.61**	31.87
R268A	35947.38	688.89	34233.23	17.80	**32580.85**	32.46	R268C	13082.88	702.95	10865.64	19.80	**10454.96**	28.00
R269A	**45496.03**	447.63	49347.98	19.33	**45496.03**	29.91	R269C	18567.27	469.55	17034.98	17.20	**16551.68**	29.67
R2610A	42371.50	437.70	42719.16	19.41	**42085.73**	32.28	R2610C	17087.31	521.34	16213.20	18.38	**15482.84**	27.36
RC161A	38981.34	723.89	38568.22	21.20	**37966.63**	20.78	RC161C	18691.07	742.28	18469.16	19.52	**17829.78**	24.06
RC162A	36931.09	701.08	36115.10	22.45	**35558.83**	27.46	RC162C	17192.16	703.69	16864.48	20.17	**16400.23**	30.90
RC163A	35118.63	694.70	34183.42	22.06	**33888.57**	36.54	RC163C	15987.36	703.31	15442.62	19.81	**15227.84**	43.97
RC164A	34036.86	714.39	33177.38	22.61	**33100.75**	42.06	RC164C	15029.87	729.91	14541.66	20.72	**14395.73**	44.37
RC165A	37890.20	698.89	37031.67	21.27	**36368.56**	26.66	RC165C	18122.45	700.42	17505.80	20.86	**17015.78**	27.02
RC166A	36917.19	688.64	36479.06	24.06	**36087.67**	24.17	RC166C	17417.86	692.80	17359.22	18.42	**16924.63**	25.93
RC167A	36045.20	673.61	35613.49	23.75	**35130.32**	27.58	RC167C	16906.27	687.48	16801.52	20.08	**16317.85**	29.50
RC168A	35731.11	658.69	34989.07	21.52	**34769.63**	32.99	RC168C	16665.67	667.88	16303.91	20.06	**15989.32**	35.12
RC169A	35355.26	661.61	34969.48	24.63	**34614.31**	32.03	RC169C	16387.88	667.16	16224.11	19.67	**15867.19**	33.63
RC1610A	35048.50	647.77	34707.66	22.63	**34405.79**	34.49	RC1610C	16380.54	666.22	15828.02	21.30	**15625.16**	37.94
RC261A	50965.70	728.20	49713.11	20.45	**48026.06**	35.41	RC261C	16887.29	517.75	16473.45	16.48	**15836.79**	27.77
RC262A	45977.09	669.75	46584.05	18.78	**43708.88**	26.18	RC262C	15607.40	611.44	14470.66	15.52	**14027.61**	28.72
RC263A	40652.52	675.24	40861.20	15.86	**38816.75**	30.87	RC263C	14175.00	716.31	12795.89	17.36	**12445.42**	25.99
RC264A	34595.77	711.36	34942.97	18.14	**33183.89**	27.61	RC264C	11806.37	749.49	10402.35	18.97	**10039.03**	24.60
RC265A	**46462.01**	552.95	47600.73	18.58	**46462.01**	31.39	RC265C	16152.60	525.72	15449.32	16.30	**14886.59**	30.69
RC266A	47640.27	589.58	47295.00	18.53	**45939.67**	27.96	RC266C	15578.28	581.92	15413.57	17.66	**14649.05**	25.77
RC267A	44631.66	655.25	**44216.27**	18.97	44922.08	28.47	RC267C	15535.85	542.03	14576.06	17.66	**13939.88**	26.18
RC268A	**41694.33**	487.59	42675.06	17.70	**41694.33**	31.84	RC268C	14289.59	536.38	14095.21	18.23	**13314.74**	26.62
RC269A	40365.67	514.88	**39103.91**	17.44	**39103.91**	25.82	RC269C	14628.42	614.84	13627.13	16.22	**12864.97**	30.58
RC2610A	38217.84	485.91	**37492.89**	18.00	40310.55	24.01	RC2610C	13745.89	578.67	13019.75	17.02	**12191.27**	22.83

(continued)

(continued)

Data set	MSDAL		Normal		New		Data set	MSDAL		Normal		New	
	Best	CPU	Best	CPU	Best	CPU		Best	CPU	Best	CPU	Best	CPU
C181A	85904.80	1331.39	85377.70	24.34	**85118.97**	27.71	C181C	31846.67	1306.95	31306.21	26.94	**31287.09**	27.80
C182A	84580.95	1330.84	83500.72	30.70	**82931.70**	41.08	C182C	31581.31	1348.14	30995.73	32.52	**30758.37**	42.99
C183A	80617.52	1296.75	80226.37	29.25	**79244.66**	54.23	C183C	29319.89	1433.02	29599.34	33.39	**28994.36**	60.33
C184A	77870.83	1309.70	77236.34	31.89	**76324.25**	55.79	C184C	27860.33	1410.50	27715.07	33.94	**27226.22**	64.46
C185A	86500.22	1329.36	85756.73	33.09	**85178.09**	28.85	C185C	31521.56	1433.78	31268.45	31.39	**31255.85**	34.35
C186A	86386.20	1319.61	86476.10	32.55	**85170.68**	37.39	C186C	31456.02	1448.78	31312.16	31.77	**31258.70**	40.11
C187A	86215.78	1350.13	86513.45	33.97	**85384.66**	41.17	C187C	31456.56	1478.83	31284.95	30.66	**31239.50**	41.04
C188A	86156.11	1319.23	85234.16	35.94	**84384.02**	46.14	C188C	31424.64	1447.25	31179.57	34.95	**30981.25**	48.24
C189A	84033.43	1313.17	83810.09	36.45	**82275.05**	49.51	C189C	30327.43	1438.94	30582.53	35.94	**29660.17**	52.90
C1810A	82053.96	1331.73	82036.56	37.22	**81018.18**	59.02	C1810C	29490.43	1418.89	29925.15	34.31	**29143.52**	59.14
C281A	81655.68	1297.78	78151.77	24.05	**76114.02**	30.62	C281C	18867.04	1433.24	19104.26	25.02	**18067.45**	30.89
C282A	79202.33	1341.41	76323.83	24.59	**75577.78**	35.96	C282C	19618.88	1445.19	19022.27	25.55	**18033.72**	38.50
C283A	78244.81	1319.11	76453.56	24.36	**75004.70**	38.39	C283C	19866.43	1533.70	18778.22	24.98	**18271.39**	44.69
C284A	74952.77	1458.69	73761.73	23.89	**73570.12**	44.88	C284C	19095.48	1621.47	18312.74	24.53	**17753.38**	44.15
C285A	80079.09	1291.39	77151.81	25.80	**75848.80**	31.31	C285C	18951.86	1323.30	19142.23	23.69	**18236.47**	30.31
C286A	78828.71	1308.94	77715.28	25.59	**75134.46**	33.04	C286C	18824.79	1250.08	18883.89	26.00	**18202.91**	33.96
C287A	80072.08	1317.11	77071.63	25.75	**75469.62**	30.23	C287C	19959.09	1278.97	19662.82	26.00	**18082.19**	36.64
C288A	78157.06	1336.98	76148.19	23.06	**75053.58**	34.74	C288C	19567.56	1278.56	18870.43	26.09	**17982.05**	31.87
C289A	78640.20	1339.33	76617.84	24.99	**75194.23**	36.74	C289C	19663.36	1302.88	19153.84	25.03	**18247.55**	34.02
C2810A	76765.70	1325.14	75010.33	25.22	**74068.15**	35.91	C2810C	19182.40	1297.06	18528.86	24.08	**17875.26**	37.60
R181A	73423.38	1262.56	72590.84	21.42	**71956.95**	38.11	R181C	40738.97	1405.81	40722.32	24.86	**39903.48**	49.58
R182A	65257.05	1202.75	64562.15	28.84	**63415.68**	48.31	R182C	35459.84	1326.39	35039.90	28.08	**34134.84**	54.85
R183A	59820.10	1182.33	57934.34	28.78	**57328.28**	54.48	R183C	30784.69	1323.24	29859.11	28.67	**29384.40**	59.03
R184A	55631.60	1226.55	54311.52	31.05	**54020.09**	65.43	R184C	27472.15	1375.56	26571.09	33.17	**26391.02**	71.90
R185A	65112.59	1170.44	64073.48	28.22	**63164.19**	36.65	R185C	36029.45	1318.28	35683.17	29.23	**34750.04**	44.60
R186A	61688.93	1162.14	59854.79	29.19	**59218.27**	49.13	R186C	32471.26	1309.08	31401.22	33.06	**31239.21**	56.55
R187A	57701.16	1170.03	56186.56	33.22	**55876.16**	54.66	R187C	29303.08	1351.52	28486.17	31.64	**28248.58**	61.46
R188A	55067.66	1240.86	54076.86	29.63	**53756.47**	70.31	R188C	27181.22	1423.31	26306.48	31.56	**26130.05**	65.15
R189A	63014.13	1158.69	61284.97	29.66	**60379.76**	39.84	R189C	33991.16	1292.28	33099.81	28.00	**32487.48**	47.46
R1810A	61051.45	1143.00	59296.93	33.11	**58539.97**	44.10	R1810C	31979.75	1272.61	31285.10	31.86	**30639.84**	49.64
R281A	76129.39	1142.08	76800.86	29.66	**76015.08**	48.08	R281C	33865.73	925.58	31588.51	26.33	**30305.13**	42.73
R282A	**68541.41**	872.50	69237.10	27.05	**68541.41**	50.33	R282C	29710.76	1113.41	27601.50	25.88	**26647.88**	42.20
R283A	60673.82	1112.23	**60335.26**	23.97	**60335.26**	51.30	R283C	25814.85	1246.00	23026.31	25.31	**22132.13**	40.90
R284A	52897.81	1199.94	**52007.50**	22.39	**52007.50**	39.33	R284C	19584.68	1263.48	18353.81	25.31	**17249.25**	39.42
R285A	**70570.70**	1043.31	73100.38	31.02	**70570.70**	44.84	R285C	31565.70	830.22	29379.08	25.31	**28067.24**	40.05
R286A	**63607.13**	844.80	66345.27	24.20	**63607.13**	50.94	R286C	28039.16	1109.69	25993.16	25.08	**24831.53**	45.52
R287A	59941.44	1180.83	**59135.41**	25.00	**59135.41**	45.30	R287C	25839.81	1042.83	22317.59	24.44	**21110.35**	41.98
R288A	50457.42	1234.20	**49862.94**	25.02	**49862.94**	42.53	R288C	19474.12	1223.72	17544.24	23.49	**16677.97**	35.86
R289A	**67169.72**	783.03	70365.53	27.66	**67169.72**	46.77	R289C	30039.87	763.03	27794.89	23.05	**26471.23**	43.24
R2810A	**64422.00**	634.05	65480.44	26.00	**64422.00**	42.81	R2810C	28481.98	883.42	26502.74	24.23	**25153.17**	37.70
RC181A	59150.52	1237.28	58036.37	28.78	**56448.46**	34.45	RC181C	32201.65	1253.56	31311.17	32.22	**30538.56**	36.15
RC182A	55364.17	1204.48	53906.74	31.30	**53097.49**	43.53	RC182C	29402.58	1234.13	28263.09	30.61	**27689.98**	47.92
RC183A	52680.52	1231.88	51423.03	31.91	**51310.77**	55.88	RC183C	27766.02	1252.24	26776.21	32.92	**26179.77**	63.58
RC184A	50474.12	1265.33	49389.16	30.49	**49314.78**	61.03	RC184C	25379.72	1285.17	24802.80	30.33	**24662.89**	55.69
RC185A	57066.52	1202.56	55811.28	29.91	**54591.68**	38.49	RC185C	30521.64	1225.08	30023.54	32.64	**29334.44**	44.85
RC186A	55904.77	1207.13	55368.05	31.22	**54173.58**	39.02	RC186C	29693.74	1213.64	29376.62	33.39	**28916.35**	44.67
RC187A	54825.86	1186.95	53951.41	31.86	**53099.90**	46.74	RC187C	29040.12	1201.38	28469.93	30.66	**27862.34**	47.72
RC188A	54077.31	1172.06	52573.27	29.80	**52058.95**	44.05	RC188C	28744.38	1194.99	27713.70	34.05	**27190.09**	53.29
RC189A	54060.53	1164.98	52536.89	31.53	**51820.68**	52.42	RC189C	28365.75	1190.58	27503.60	30.81	**27073.12**	59.78
RC1810A	52977.40	1158.20	52055.42	30.16	**51203.98**	55.18	RC1810C	27446.71	1191.06	26982.44	31.91	**26730.07**	56.44
RC281A	72564.38	1180.42	**68932.23**	29.66	**68932.23**	46.29	RC281C	26967.79	1094.53	25510.42	25.92	**24403.06**	39.81
RC282A	65076.57	1235.95	**63429.73**	23.53	**63429.73**	52.39	RC282C	24967.48	1241.05	22564.77	23.78	**21796.86**	41.23
RC283A	57898.17	1325.23	**56750.79**	23.84	**56750.79**	46.47	RC283C	21165.76	1392.27	19486.10	24.56	**18750.79**	46.96
RC284A	51738.96	1349.83	49084.96	22.77	**48176.69**	40.01	RC284C	17256.03	1489.83	**15650.35**	24.39	16048.69	35.16
RC285A	**65302.06**	1051.83	65449.16	25.58	**65302.06**	57.86	RC285C	24480.05	1150.34	23440.24	24.03	**23403.65**	43.96
RC286A	65842.89	1086.66	**62731.40**	23.25	**62731.40**	54.91	RC286C	24864.24	1038.98	23389.50	24.44	**22937.89**	40.26
RC287A	63694.95	1030.31	**61200.18**	25.38	**61200.18**	51.48	RC287C	23640.37	1088.58	22536.80	27.53	**22031.78**	44.43
RC288A	59340.22	977.31	**57763.84**	22.97	**57763.84**	52.43	RC288C	22731.98	1052.16	21481.50	24.59	**20758.74**	36.79
RC289A	57960.22	995.58	**56770.02**	23.45	**56770.02**	49.76	RC289C	22423.33	1115.30	20964.52	23.45	**20641.88**	37.77
RC2810A	54409.41	998.69	**54190.70**	22.80	**54190.70**	42.84	RC2810C	21656.00	1091.56	20208.70	25.58	**19503.18**	35.71

(continued)

(continued)

Data set	MSDAL		Normal		New		Data set	MSDAL		Normal		New	
	Best	CPU	Best	CPU	Best	CPU		Best	CPU	Best	CPU	Best	CPU
C1101A	118787.14	1983.70	117485.38	35.27	**117207.95**	35.07	C1101C	50787.86	2021.80	50065.08	36.55	**50057.53**	36.38
C1102A	116725.88	1987.66	115410.43	46.91	**114115.83**	58.16	C1102C	49784.06	2048.88	49606.12	45.03	**49365.50**	64.76
C1103A	112483.80	1981.67	111346.16	47.77	**109840.98**	65.74	C1103C	47210.25	2058.17	47948.33	49.47	**47078.95**	73.02
C1104A	106391.73	2042.83	104794.52	44.94	**104186.72**	83.97	C1104C	43981.17	2177.30	43895.81	48.11	**42999.37**	83.60
C1105A	118762.64	2028.78	117881.98	42.61	**117198.64**	43.28	C1105C	50310.32	2201.94	50075.16	45.20	**50032.87**	45.99
C1106A	120062.98	2027.02	118269.16	46.00	**117143.79**	51.76	C1106C	50139.08	2208.06	50131.77	42.36	**50033.15**	48.81
C1107A	118882.44	2044.41	119287.70	43.56	**117212.94**	47.02	C1107C	50344.85	2201.61	50179.69	43.84	**50023.45**	51.35
C1108A	118720.60	2010.92	117465.45	47.41	**116276.77**	57.52	C1108C	50059.20	2206.91	50078.75	44.41	**49692.13**	65.60
C1109A	115666.78	2043.36	114729.25	49.77	**114032.42**	70.51	C1109C	48281.63	2239.78	49033.63	45.05	**47616.00**	69.56
C11010A	113723.05	2018.86	113565.70	48.25	**111928.58**	70.01	C11010C	47767.31	2189.56	48006.15	48.13	**46606.41**	82.70
C2101A	104405.41	1895.28	101055.51	33.80	**99171.59**	41.22	C2101C	27565.94	2273.83	26914.42	33.97	**25081.81**	38.61
C2102A	104122.18	1910.50	**99865.72**	35.19	**99865.72**	45.19	C2102C	27451.23	2237.17	26360.07	34.03	**24946.05**	52.63
C2103A	100344.77	1969.92	**97243.95**	32.99	**97243.95**	52.12	C2103C	27775.83	2340.63	26335.14	33.88	**25294.95**	51.11
C2104A	99061.77	2187.19	95365.89	32.52	**95303.05**	60.72	C2104C	26171.65	2457.75	25404.70	35.55	**24720.91**	64.02
C2105A	102803.17	1850.39	**99935.95**	34.17	**99935.95**	42.62	C2105C	27069.58	2163.28	26709.19	35.14	**25164.18**	44.40
C2106A	102681.81	1896.06	99364.25	32.88	**97425.92**	43.43	C2106C	27874.57	2252.98	26926.17	31.20	**24826.73**	47.72
C2107A	103154.60	1881.92	99228.42	33.66	**98391.09**	47.24	C2107C	27906.83	2217.74	26664.01	32.72	**26200.46**	51.23
C2108A	101995.64	1915.75	97999.53	34.11	**96851.00**	42.84	C2108C	27492.01	2220.94	26398.24	34.13	**25584.10**	47.13
C2109A	101620.37	1898.66	98829.27	33.20	**97843.37**	46.50	C2109C	27589.01	2241.56	26563.02	33.48	**25962.44**	51.42
C21010A	101197.13	1938.42	96876.09	32.73	**96513.50**	46.11	C21010C	26891.55	2240.75	26151.47	32.24	**24881.47**	50.65
R1101A	100212.48	2040.98	98182.30	30.14	**97552.74**	49.45	R1101C	58943.11	2178.30	58614.10	28.39	**57340.91**	72.46
R1102A	89838.07	1853.09	87719.66	39.27	**86698.79**	66.94	R1102C	51655.98	2084.27	50395.03	42.08	**49488.02**	75.13
R1103A	81944.28	1725.16	79314.20	40.33	**78694.48**	76.91	R1103C	45856.57	2062.03	43793.72	40.02	**43168.66**	90.37
R1104A	76926.73	1806.22	74883.88	40.28	**74448.69**	92.68	R1104C	41019.65	2139.47	39729.29	41.39	**39336.09**	95.64
R1105A	90778.97	1703.67	87170.08	36.02	**86018.13**	55.86	R1105C	53446.32	2078.80	51998.27	40.47	**50455.79**	65.24
R1106A	85998.64	1694.72	82132.25	39.27	**81654.91**	67.05	R1106C	48283.29	2045.13	46956.65	42.64	**45874.48**	81.85
R1107A	80195.64	1733.50	77698.93	40.56	**77295.04**	86.99	R1107C	44417.54	2087.34	42431.94	40.38	**41982.23**	91.40
R1108A	76503.94	1761.13	74621.00	43.72	**74191.95**	90.26	R1108C	40368.48	2159.92	39378.73	44.86	**39012.00**	95.35
R1109A	88163.81	1688.74	85080.58	37.31	**83994.10**	59.11	R1109C	50974.68	2044.73	49370.69	41.14	**48257.65**	69.27
R11010A	85534.08	1671.95	82393.37	39.45	**81317.50**	64.04	R11010C	48615.96	2017.91	46699.48	42.25	**45601.70**	72.32
R2101A	110184.33	1313.19	104301.16	42.06	**101743.42**	61.65	R2101C	49601.56	1829.30	45786.67	35.08	**43912.34**	58.24
R2102A	97442.41	1501.20	94750.47	38.34	**91999.25**	68.70	R2102C	45550.71	1918.45	39158.92	33.06	**38138.85**	65.75
R2103A	83346.44	1714.33	80347.20	38.19	**79405.92**	67.33	R2103C	37908.70	2233.11	31949.84	30.06	**30898.94**	55.55
R2104A	70656.05	1757.58	**65589.11**	28.13	65616.95	58.11	R2104C	28511.63	2028.28	24545.11	33.19	**23218.64**	58.45
R2105A	104693.34	1206.02	**100037.95**	42.27	**100037.95**	63.35	R2105C	46757.77	1514.56	43108.04	34.14	**40929.15**	57.30
R2106A	90800.22	1400.36	**87520.57**	35.44	**87520.57**	53.21	R2106C	42134.09	1608.91	37719.80	30.58	**35894.18**	57.13
R2107A	76471.25	1706.22	**75569.19**	34.63	**75569.19**	63.94	R2107C	35172.42	1887.48	30835.81	33.17	**29276.42**	62.03
R2108A	69906.48	1778.36	**63537.16**	31.73	**63537.16**	55.55	R2108C	27657.81	2122.02	23537.39	32.70	**22319.10**	50.09
R2109A	**96569.63**	1050.55	97251.31	40.09	**96569.63**	61.37	R2109C	44953.75	1488.67	40990.00	34.45	**38742.40**	56.63
R21010A	**90303.81**	1071.19	93063.65	37.48	**90303.81**	60.25	R21010C	43142.36	1199.31	39200.86	35.94	**38064.95**	48.47
RC1101A	82815.73	1876.09	79959.62	39.94	**78031.41**	48.69	RC1101C	48849.41	2037.16	47049.28	39.42	**45756.90**	49.30
RC1102A	77839.84	1885.22	73959.01	42.27	**72774.02**	53.13	RC1102C	44660.72	2122.11	42589.15	42.98	**41655.57**	65.68
RC1103A	72778.05	1865.02	70309.67	44.02	**69879.53**	78.53	RC1103C	41370.69	2157.48	39278.03	41.47	**38931.45**	82.88
RC1104A	69462.56	1937.08	68081.25	41.99	**67776.59**	92.87	RC1104C	38606.30	2212.52	37534.31	43.00	**37178.45**	100.22
RC1105A	79230.56	1877.63	76776.20	40.53	**75250.41**	54.88	RC1105C	46503.62	2113.03	44786.32	40.30	**43533.30**	57.16
RC1106A	77781.09	1877.72	75310.20	42.14	**73961.84**	51.04	RC1106C	46316.18	2081.97	44116.07	40.05	**42585.26**	44.74
RC1107A	75976.66	1845.33	73748.70	44.55	**72905.91**	58.67	RC1107C	45117.43	2072.89	43211.56	43.08	**42054.25**	55.51
RC1108A	75322.36	1837.09	72488.42	40.33	**71715.46**	60.39	RC1108C	44002.29	2058.11	41422.41	40.25	**40985.61**	59.87
RC1109A	74556.65	1833.77	72499.64	40.98	**71628.86**	54.88	RC1109C	43665.91	2002.34	41758.04	40.00	**40977.04**	57.14
RC11010A	74562.46	1826.00	71752.19	41.52	**70987.31**	50.56	RC11010C	42647.78	1875.02	40857.08	39.42	**40109.00**	61.37
RC2101A	96431.08	1718.86	**92100.95**	36.59	**92100.95**	53.59	RC2101C	39684.66	1405.31	36591.16	36.44	**34879.15**	47.97
RC2102A	87586.31	1651.05	**82790.18**	33.86	**82790.18**	44.52	RC2102C	36250.77	1799.69	32209.94	31.31	**30925.31**	58.36
RC2103A	74445.37	1918.56	**69548.62**	25.39	**69548.62**	48.25	RC2103C	30443.51	2185.44	26761.99	32.25	**25570.40**	56.79
RC2104A	64687.95	2113.05	**61890.14**	33.45	**61890.14**	49.84	RC2104C	25704.52	2427.97	21596.82	36.56	**20690.84**	49.12
RC2105A	88789.04	1470.70	**87244.16**	36.31	**87244.16**	49.94	RC2105C	36891.79	1587.11	34349.43	36.20	**32316.64**	48.74
RC2106A	90015.29	1518.77	**85087.14**	35.89	**85087.14**	44.62	RC2106C	37837.96	1567.25	34312.05	32.77	**32194.44**	45.96
RC2107A	86259.89	1397.72	**83513.57**	34.17	**83513.57**	47.02	RC2107C	36658.46	1595.92	33036.03	33.45	**30571.31**	51.40
RC2108A	**80197.05**	1495.63	80933.86	31.77	**80197.05**	57.44	RC2108C	33952.09	1539.23	31815.12	33.99	**29466.26**	42.56
RC2109A	79337.48	1414.70	**73604.45**	33.33	**73604.45**	44.85	RC2109C	33757.40	1591.59	30717.02	31.81	**28540.66**	49.22
RC21010A	75617.04	1443.11	**71194.46**	33.72	**71194.46**	45.83	RC21010C	33072.73	1567.84	29522.06	34.28	**27463.52**	49.70

References

Baldacci R, Battarra M, Vigo D (2008) Routing a heterogeneous fleet of vehicles. In: The vehicle routing problem: latest advances and new challenges. Springer, Heidelberg, pp 3–27

Bräysy O (2003) A reactive variable neighborhood search for the vehicle-routing problem with time windows. INFORMS J Comput 15(4):347–368

Bräysy O, Dullaert W, Hasle G, Mester D, Gendreau M (2008) An effective multirestart deterministic annealing metaheuristic for the fleet size and mix vehicle-routing problem with time windows. Transp Sci 42(3):371–386

Bräysy O, Porkka PP, Dullaert W, Repoussis PP, Tarantilis CD (2009) A well-scalable metaheuristic for the fleet size and mix vehicle routing problem with time windows. Expert Syst Appl 36(4):8460–8475

Campbell AM, Savelsbergh M (2004) Efficient insertion heuristics for vehicle routing and scheduling problems. Transp sci 38(3):369–378

Clarke GU, Wright JW (1964) Scheduling of vehicles from a central depot to a number of delivery points. Oper Res 12(4):568–581

Gehring H, Homberger J (1999) A parallel hybrid evolutionary metaheuristic for the vehicle routing problem with time windows. In: Proceedings of EUROGEN99, vol. 2. Citeseer

Gendreau M, Potvin J-Y, Bräumlaysy O, Hasle G, Løkketangen A (2008) Metaheuristics for the vehicle routing problem and its extensions: a categorized bibliography. In: The vehicle routing problem: latest advances and new challenges. Springer, Heidelberg, pp 143–169

Kirkpatrick S, Gelatt CD, Vecchi MP et al (1983) Optimization by simmulated annealing. Science 220(4598):671–680

Kytöjoki J, Nuortio T, Bräysy O, Gendreau M (2007) An efficient variable neighborhood search heuristic for very large scale vehicle routing problems. Comput Oper Res 34(9):2743–2757

Liu F-H, Shen S-Y (1999) The fleet size and mix vehicle routing problem with time windows. J Oper Res Soc 50(7):721–732

Or I (1976) Traveling salesman-type combinatorial problems and their relation to the logistics of regional blood banking. Xerox University Microfilms

Solomon MM (1987) Algorithms for the vehicle routing and scheduling problems with time window constraints. Oper Res 35(2):254–265

Chapter 5
Green Route Allocation in a Transportation Network

Victor Zakharov, Alexander Krylatov and Dmitriy Volf

Abstract A lack of methodological tools that can be used to support decision makers in decreasing greenhouse gas emission levels has motivated us to write this paper. This paper investigates the problem of determining the allocation of available green route capacity. The approach for designing a green transit network that offers green vehicles shorter travel times between given origins and destinations is discussed. This approach is extended to minimize greenhouse gas emissions. For this purpose, bi-level programs are formulated to minimize the emission function under competitive and non-competitive scenarios.

5.1 Introduction

It is well-known that large amount of traffic flow is one of the main contributors of air pollution. The authorities in big cities are interested in decreasing the level of emissions, but the implementation of appropriate arrangements requires developed methodological tools. Therefore, it is highly important to investigate models and methods to address the problem of greenhouse gas emissions. Many researchers work in different directions by studying the interdependence between the level of emissions and travel decisions, including choosing a route, the departure time, and the mode of transport (Ben-Akiva et al. 1991; Gaker et al. 2011; Mahmassani 1990). These directions are especially remarkable because routes that minimize travel time

V. Zakharov (✉) · A. Krylatov · D. Volf
Saint-Petersburg State University, Saint-Petersburg, Russia
e-mail: mcvictor@mail.ru

A. Krylatov
e-mail: aykrylatov@yandex.ru

D. Volf
e-mail: answer.iii@mail.ru

P. Diez et al. (eds.), *Computational Methods and Models for Transport*,
Computational Methods in Applied Sciences 45,
DOI 10.1007/978-3-319-54490-8_5

are not the same as routes that minimize emissions (Ahn and Rakha 2008; Lin and Ge 2006). The influence of physical road parameters on the state of pollution has also been studied widely (Boriboonsomsin and Barth 2009; Boroujeni and Frey 2014; Wyatt et al. 2014; Zhang and Frey 2006). The problem of green (vehicles with low emission levels) and non-green vehicle assignment is deeply investigated in Zakharov and Krylatov (2015a). According to Zakharov and Krylatov (2015a, b), authorities of large cities are typically interested in encouraging the use of environmentally friendly vehicles in transportation networks because green vehicles decrease the total greenhouse gas emissions. Thus, appropriate arrangements should be employed to motivate drivers to use green vehicles instead of gasoline-powered vehicles. To achieve this goal, a transit network designed for green-vehicle routing is an effective method (Zakharov and Krylatov 2015b; Beltran et al. 2009). However, it should be noted that if we provide too many routes to green vehicles, then the traffic on saturated non-green routes could lead to increases in emissions. Therefore, this paper addresses the problem of green route allocation under two requirements: (1) green routes must be attractive for drivers (less travel time) and (2) green route allocation must lead to the minimal emission levels.

The paper is organized as follows. Section 5.2 introduces a generalized formulation of the green and non-green traffic assignment problem under the user-equilibrium principle. In Sect. 5.3 we discuss the problem of optimal green route allocation to minimize greenhouse gas emissions. Conclusions are briefly summarized in Sect. 5.4.

5.2 Optimal Green Route Allocation in the Case of User-Equilibrium Assignment

A user-equilibrium assignment of both green and non-green vehicles on a network with parallel routes was investigated in Zakharov and Krylatov (2015a). In this section, first, we will discuss key results that we already obtained in this area, and, second, we will generalize some of our previous mathematical formulations.

According to Zakharov and Krylatov (2015a, b), green routes are available for green vehicles only and should be allocated *optimal*, i.e., in a user-equilibrium state, the travel time of green vehicles is less than or equal to the travel time of non-green vehicles between the same origin-destination nodes (Zakharov and Krylatov 2014b). Moreover, we developed a procedure for defining this set of green routes where the optimal allocation takes place. For this purpose, we consider the problems with the traffic equilibrium assignment on the green and non-green subnetworks separately. First, we introduced a transportation network with an oriented graph $G = (N, A)$, where W is the set of origin-destination pairs of nodes (OD-pairs), $w \in W$. Then, we divided the set of all routes K^w into two sets: the set of green routes K_1^w and the set of non-green routes K_2^w. Therefore, this allowed us to formulate two independent optimization programs:

(A) For green routes

$$\min_g Z_1 = \min_g \sum_a \int_0^{x_a} t_a(u)du,$$

with constraints $\forall\, w \in W$

$$\sum_{k \in K_1^w} g_k = G^w,$$

$$g_k \geq 0, \quad \forall k \in K_1^w,$$

with definitional constraint

$$x_a = \sum_{w \in W} \sum_{k \in K_1^w} g_k \delta_{a,k}^w, \quad \forall a \in A,$$

(B) For non-green routes

$$\min_f Z_2 = \min_f \sum_a \int_0^{x_a} t_a(u)du,$$

with constraints $\forall\, w \in W$

$$\sum_{k \in K_2^w} f_k = F^w,$$

$$f_k \geq 0, \quad \forall k \in K_2^w,$$

with definitional constraint

$$x_a = \sum_{w \in W} \sum_{k \in K_2^w} f_k \delta_{a,k}^w, \quad \forall a \in A,$$

where G^w—demand of green vehicles between OD-pair w; F^w—demand of non-green vehicles between OD-pair w; g_k—the flow of green vehicles on route k, $k \in K_1^w$, $g = \{g_k\}_{k \in K_1^w}^{w \in W}$; f_k—the flow of green and non-green vehicles on route k, $k \in K_2^w$, $f = \{f_k\}_{k \in K_2^w}^{w \in W}$; x_a—traffic flow on the arc $a \in A$; $t_a(x_a)$—travel time of flow x_a through congested arc $a \in A$; $\delta_{a,k}^w$—indicator: 1 if arc a "belongs" to the route k, and 0 otherwise.

Optimization programs A and B lead to a user-equilibrium assignment on the green and non-green subnetworks, respectively, when green vehicles use only green routes (Sheffi 1985). Clearly, for green vehicles, it is reasonable to drive through green routes only if their travel times are less than or equal to the travel times of non-green vehicles. Therefore, the sets of green routes K_1^w $\forall\, w \in W$ are optimal if and only if the optimal travel time of green vehicles from program A is less than or equal to the optimal travel time of non-green vehicles from program B.

Now, let us introduce the additional notation: G_1^w—the flow of green vehicles through green routes between OD-pair w, $G_1 = \{G_1^w\}_{w \in W}$; G_2^w—the flow of green and non-green vehicles through non-green routes between OD-pair w, $G_2 = \{G_2^w\}_{w \in W}$. Then, the statement for the following theorem occurs.

Theorem 1 *User-equilibrium assignment of green and non-green vehicles on the general network with green routes could be found as a solution of the following optimization program:*

$$\min_{x,G_1,G_2} Z(x, G_1, G_2) = \min_{x,G_1,G_2} \sum_{a \in A} \int_0^{x_a} t_a(u)du, \tag{5.1}$$

with constraints $\forall\, w \in W$

$$\sum_{k \in K_1^w} g_k = G_1^w, \tag{5.2}$$

$$\sum_{k \in K_2^w} f_k = G_2^w + F^w, \tag{5.3}$$

$$G_1^w + G_2^w = G^w, \tag{5.4}$$

$$g_k \geq 0 \quad \forall k \in K_1^w, \tag{5.5}$$

$$f_k \geq 0 \quad \forall k \in K_2^w, \tag{5.6}$$

$$G_1^w \geq 0 \quad and \quad G_2^w \geq 0, \tag{5.7}$$

with definitional constraint

$$x_a = \sum_w \sum_{k \in K_1^w} g_k \delta_{a,k}^w + \sum_w \sum_{k \in K_2^w} f_k \delta_{a,k}^w, \quad \forall a \in A. \tag{5.8}$$

Proof By definition, user-equilibrium corresponds to the situation when the journey times in all routes actually used are equal to and less than those that would be experienced by a single vehicle on any unused route.

The Lagrangian of problem (5.1)–(5.7) is

$$L = \sum_{a \in A} \int_0^{x_a} t_a(u)du + \sum_{w \in W} \Bigg[\omega^w \Bigg(G_1^w - \sum_{k \in K_1^w} g_k \Bigg) +$$

$$+ \nu^w \Bigg(G_2^w + F^w - \sum_{k \in K_2^w} f_k \Bigg) + \sigma^w \Big(G^w - G_1^w - G_2^w \Big) +$$

$$+\sum_{w\in W}\sum_{k\in K_1^w}(-g_k)\eta_k^w+\sum_{w\in W}\sum_{k\in K_2^w}(-f_k)\xi_k^w+$$

$$\left.+\gamma_1^w(-G_1^w)+\gamma_2^w(-G_2^w)\right],$$

where ω^w, ν^w, $\eta_k^w \geq 0$ $(k \in K_1^w)$, $\xi_k^w \geq 0$ $(k \in K_2^w)$, γ_1^w and γ_2^w $\forall\ w \in W$ are Lagrangian multipliers.

Goal functions are convex, and Kuhn-Tucker conditions hold. Differentiate L by all unknown variables and equate it to zero $\forall\ w \in W$:

$$\frac{\partial L}{\partial g_k}=\frac{\partial Z}{\partial g_k}-\omega^w-\eta_k^w=0,\quad k\in K_1^w, \tag{5.9}$$

$$\frac{\partial L}{\partial f_k}=\frac{\partial Z}{\partial f_k}-\nu^w-\xi_k^w=0,\quad k\in K_2^w, \tag{5.10}$$

$$\frac{\partial L}{\partial G_1^w}=\omega^w-\sigma^w-\gamma_1^w=0, \tag{5.11}$$

$$\frac{\partial L}{\partial G_2^w}=\nu^w-\sigma^w-\gamma_2^w=0. \tag{5.12}$$

Note that for $k \in K_1^w$ $\forall\ w \in W$:

$$\frac{\partial Z}{\partial g_k}=\sum_{a\in A}\frac{\partial Z}{\partial x_a}\frac{\partial x_a}{\partial g_k}=\sum_{a\in A}t_a(x_a)\delta_{a,k}^w=T_k(g_k),$$

and for $k \in K_2^w$ $\forall\ w \in W$:

$$\frac{\partial Z}{\partial f_k}=\sum_{a\in A}\frac{\partial Z}{\partial x_a}\frac{\partial x_a}{\partial f_k}=\sum_{a\in A}t_a(x_a)\delta_{a,k}^w=T_k(f_k).$$

Moreover, for conditions (5.5)–(5.7) complementary slackness holds and, consequently, (5.9)–(5.12) could be reformulated $\forall\ w \in W$

$$T_k(g_k)\begin{cases}=\omega^w \text{ for } g_k>0,\\ \geq \omega^w \text{ for } g_k=0,\end{cases}\quad k\in K_1^w, \tag{5.13}$$

$$T_k(f_k)\begin{cases}=\nu^w \text{ for } f_k>0,\\ \geq \nu^w \text{ for } f_k=0,\end{cases}\quad k\in K_2^w, \tag{5.14}$$

$$\omega^w \begin{cases} = \sigma^w \text{ for } G_1^w > 0, \\ \geq \sigma^w \text{ for } G_1^w = 0, \end{cases} \tag{5.15}$$

$$\nu^w \begin{cases} = \sigma^w \text{ for } G_2^w > 0, \\ \geq \sigma^w \text{ for } G_2^w = 0. \end{cases} \tag{5.16}$$

Therefore, conditions (5.13) and (5.14) guarantee the user-equilibrium assignment on the green and non-green subnetworks, respectively. Moreover, if $G_1^w > 0$ and $G_2^w > 0$, then $\omega^w = \sigma^w$ and $\nu^w = \sigma^w$. Hence, $\omega^w = \nu^w$. In this case, $T_k(g_k) = T_{\mathfrak{k}}(f_{\mathfrak{k}}), k \in K_1^w, \mathfrak{k} \in K_2^w$. Therefore, the user-equilibrium assignment is on the whole network.

Theorem is proved.

Due to the proof of Theorem 1, we obtain $\forall\, w \in W$. If $G_1^w = G^w$ and $G_2^w = 0$, then $\omega^w = \sigma^w$ and $\nu^w \geq \sigma^w$. Hence, $\omega^w \leq \nu^w$ and, consequently,

$$T_k(g_k) \leq T_{\mathfrak{k}}(f_{\mathfrak{k}}), \quad k \in K_1^w, \quad \mathfrak{k} \in K_2^w. \tag{5.17}$$

Therefore, instead of using a procedure based on optimization programs A and B, we can use a procedure based on the optimization program from Theorem 1. Actually, according to (5.17), the sets of green routes $K_1^w\ \forall\, w \in W$ are optimal if and only if the optimal solution (g^*, f^*, G_1^*, G_2^*) of program (5.1)–(5.8) is such, that $G_1^{w*} = G^w$ and $G_2^{w*} = 0\ \forall\, w \in W$.

It is remarkable that for a network of parallel routes presented by digraph with one origin-destination pair and n parallel links where each link is associated with a route from origin to destination, we obtained explicit conditions to define an optimal green route allocation:

Theorem 2 *n_1 is optimal if and only if the following two conditions hold:*

$$G > \sum_{i=1}^{n_1} c_i \left(\frac{t_{n_1}^0}{t_i^0} - 1 \right) \quad \textit{and} \quad F > \sum_{i=n_1+1}^{n_2} c_i \left(\frac{t_{n_2}^0}{t_i^0} - 1 \right), \tag{5.18}$$

$$\frac{G + \sum_{i=1}^{n_1} c_i}{\sum_{i=1}^{n_1} \frac{c_i}{t_i^0}} \leq \frac{F + \sum_{i=n_1+1}^{n_2} c_i}{\sum_{i=n_1+1}^{n_2} \frac{c_i}{t_i^0}}, \tag{5.19}$$

when, without loss of generality, the routes are numbered as follows

$$t_1^0 \leq \ldots \leq t_{n_1}^0 \quad \textit{and} \quad t_{n_1+1}^0 \leq \ldots \leq t_{n_2}^0. \tag{5.20}$$

In Theorem 2, the following notation is used: $N = \{1, \ldots, n\}$—the set of routes; $N_1 = \{1, \ldots, n_1\}$—the set of green routes; $N_2 = \{n_1 + 1, \ldots, n_2 = n\}$—the set of non-green routes; G—demand of green vehicles between origin and destination;

F—number of non-green vehicles on the network; g_i—the flow of green vehicles on route i, $i = \overline{1, n_1}$, $g = (g_1, \ldots, g_{n_1})$; f_i—the flow of green and non-green vehicles on route i, $i = \overline{n_1 + 1, n_2}$, $f = (f_{n_1+1}, \ldots, f_{n_2})$; t_i^0—free travel time on route i, $i = \overline{1, n}$; c_i—capacity of route i, $i = \overline{1, n}$; $t_i(\mathfrak{F}_i) = t_i^0 \left(1 + \frac{\mathfrak{F}_i}{c_i}\right)$—travel time of flow $\mathfrak{F}_i$ through congested route i, $i = \overline{1, n}$. Travel time is modelled by the linear BPR-delay function (US Bureau of Public Roads 1964). Nonlinear BPR-function for the network of parallel routes is deeply studied in Krylatov (2016).

If G_1 is the number of green vehicles using green routes and if G_2 is the number of green vehicles using non-green routes, then due to the Theorem 1 we could claim that conditions (5.18)–(5.20) would also appear in an optimal decision for the following optimization program:

$$\min_{g, f, G_1, G_2} z(g, f, G_1, G_2) = \min_{g, f, G_1, G_2} \left[\sum_{i=1}^{n_1} \int_0^{g_i} t_i(u)du + \sum_{i=n_1+1}^{n_2} \int_0^{f_i} t_i(u)du \right],$$

with constraints

$$\sum_{i=1}^{n_1} g_i = G_1,$$

$$\sum_{i=n_1+1}^{n_2} f_i = G_2 + F,$$

$$G_1 + G_2 = G,$$

$$g_i \geq 0 \quad \forall i = \overline{1, n},$$

$$f_i \geq 0 \quad \forall i = \overline{n_1 + 1, n_2},$$

$$G_1 \geq 0 \quad \text{and} \quad G_2 \geq 0,$$

when $G_1^* = G$ and $G_2^* = 0$.

5.3 The Level of Emissions

The commonly used method to estimate emission impact is the Gaussian model (Lin and Ge 2006; Zhang et al. 2010). Note that the Gaussian model could be employed to model the emission concentration in the study area. We consider link flows as a main source of emissions and, hence, we use adopted the Gaussian function (Yin and Lawphongpanich 2006). Therefore, to estimate the level of emissions on the green routes, we will use the following function:

$$E_k(g_k) = \alpha_k \cdot t_k(g_k) \cdot e^{\beta_k \frac{l_k}{t_k(g_k)}}, \quad \forall k \in K_1^w. \tag{5.21}$$

To estimate level of emission on the non-green routes, we will use the following function:

$$E_k(g_k, f_k) = \alpha_k \cdot t_k(g_k, f_k) \cdot e^{\beta_k \frac{l_k}{t_k(g_k, f_k)}} \cdot \gamma_k, \quad \forall k \in K_2^w, \tag{5.22}$$

where

$$\gamma_k = \left(e^{\frac{g_k}{f_k}}\right)^{-1} \quad \forall k \in K_2^w. \tag{5.23}$$

It is clear, that when $g_k = 0$ for some $k \in K_2^w$, expression (5.22) has the same form as (5.21). Moreover, according to (5.22), the more green vehicles compared with non-green vehicles that use route $k \in K_2^w$, the smaller the level of emissions that could be fixed on this route. This approach is quite natural and reasonable.

To minimize the level of emissions, a decision maker could influence the set of green routes K_1^w, $w \in W$. In this case, the problem of optimal green route allocation to minimize greenhouse gas emissions could be formulated as the following optimization program:

$$\begin{gathered} \min_{K_1^w} E(g, f) = \\ \min_{K_1^w} \sum_{w \in W} \Bigg[\sum_{k \in K_1^w} t_k(g_k) \cdot e^{\frac{l_k}{t_k(g_k)}} + \\ \sum_{k \in K_2^w} t_k(g_k, f_k) \cdot e^{\frac{l_k}{t_k(g_k, f_k)}} \cdot \gamma_k \Bigg], \end{gathered} \tag{5.24}$$

when traffic flows are assigned in a user-equilibrium way:

$$\min_{x, G_1, G_2} Z(x, G_1, G_2) = \min_{x, G_1, G_2} \sum_{a \in A} \int_0^{x_a} t_a(u) du, \tag{5.25}$$

with constraints $\forall\, w \in W$

$$\sum_{k \in K_1^w} g_k = G_1^w, \tag{5.26}$$

$$\sum_{k \in K_2^w} f_k = G_2^w + F^w, \tag{5.27}$$

$$G_1^w + G_2^w = G^w, \tag{5.28}$$

$$g_k \geq 0 \;\; \forall k \in K_1^w, \tag{5.29}$$

$$f_k \geq 0 \;\; \forall k \in K_2^w, \tag{5.30}$$

$$G_1^w \geq 0 \;\text{ and }\; G_2^w \geq 0, \tag{5.31}$$

with definitional constraint

$$x_a = \sum_{w} \sum_{k \in K_1^w} g_k \delta_{a,k}^w + \sum_{w} \sum_{k \in K_2^w} f_k \delta_{a,k}^w, \quad \forall a \in A. \tag{5.32}$$

Bi-level program (5.24)–(5.32) corresponds to the situation when it is believed that all vehicles chose the route independently, i.e., the modelling of the traffic flow assignment could be based on the user-equilibrium principle. However, if it is assumed that any user of the network belongs to some UG and this UG chooses the routes for all of its users, then the modelling of the traffic flow assignment could be based on the competitive principle. Therefore, in a competitive case, the problem of optimal green route allocation to minimize greenhouse gas emissions could be formulated as the following optimization program:

$$\begin{aligned} &\min_{K_1^w} E(g, f) = \\ &\min_{K_1^w} \sum_{w \in W} \left[\sum_{k \in K_1^w} t_k(g_k) \cdot e^{\frac{l_k}{t_k(g_k)}}, \right. \\ &\left. + \sum_{k \in K_2^w} t_k(g_k, f_k) \cdot e^{\frac{l_k}{t_k(g_k, f_k)}} \cdot \gamma_k \right] \end{aligned} \tag{5.33}$$

when the traffic flows of many UGs are assigned in a system optimum way $\forall\, j \in M$ (Krylatov and Zakharov 2016):

$$\min_{x^j, G_1^j, G_2^j} Z^j(x^j, G_1^j, G_2^j) = \min_{x^j, G_1^j, G_2^j} \sum_{a \in A} t_a(x_a) x_a^j, \tag{5.34}$$

with constraints $\forall\, w \in W$

$$\sum_{k \in K_1^w} g_k^j = G_1^{j,w}, \tag{5.35}$$

$$\sum_{k \in K_2^w} f_k^j = G_2^{j,w} + F^{j,w}, \tag{5.36}$$

$$G_1^{j,w} + G_2^{j,w} = G^{j,w}, \tag{5.37}$$

$$g_k^j \geq 0 \quad \forall k \in K_1^w, \tag{5.38}$$

$$f_k^j \geq 0 \quad \forall k \in K_2^w, \tag{5.39}$$

$$G_1^{j,w} \geq 0 \;\text{ and }\; G_2^{j,w} \geq 0, \tag{5.40}$$

with definitional constraints

$$x_a^j = \sum_w \sum_{k \in K_1^w} g_k^j \delta_{a,k}^w + \sum_w \sum_{k \in K_2^w} f_k^j \delta_{a,k}^w, \quad \forall a \in A. \tag{5.41}$$

$$x_a = \sum_{j=1}^{m} x_a^j, \quad \forall j \in M. \tag{5.42}$$

It is remarkable that Theorem 2 allows us to express explicit conditions for optimal green route allocation for the problem of low level for a network of parallel routes.

Theorem 3 *The number of green routes n_1^* guarantees a minimal level of emissions on the network of parallel routes in the user-equilibrium case if and only if it is the decision of the following optimization program:*

$$\begin{gathered} \min_{n_1} E(g, f) = \\ \min_{n_1} \Big[\textstyle\sum_{i=1}^{n_1} \alpha_i \cdot t_i(g_i) \cdot e^{\beta_i \frac{l_i}{t_i(g_i)}} +, \\ \textstyle\sum_{i=n_1+1}^{n_2} \alpha_i \cdot t_i(f_i) \cdot e^{\beta_i \frac{l_i}{t_i(f_i)}} \Big] \end{gathered} \tag{5.43}$$

with constraints:

$$G > \sum_{i=1}^{n_1} c_i \left(\frac{t_{n_1}^0}{t_i^0} - 1 \right) \quad and \quad F > \sum_{i=n_1+1}^{n_2} c_i \left(\frac{t_{n_2}^0}{t_i^0} - 1 \right), \tag{5.44}$$

$$\frac{G + \sum_{i=1}^{n_1} c_i}{\sum_{i=1}^{n_1} \frac{c_i}{t_i^0}} \le \frac{F + \sum_{i=n_1+1}^{n_2} c_i}{\sum_{i=n_1+1}^{n_2} \frac{c_i}{t_i^0}}, \tag{5.45}$$

and when the routes are numbered as follows

$$t_1^0 \le \ldots \le t_{n_1}^0 \quad and \quad t_{n_1+1}^0 \le \ldots \le t_{n_2}^0, \tag{5.46}$$

where

$$g_i = \frac{c_i}{t_i^0} \frac{G + \sum_{s=1}^{n_1} c_s}{\sum_{s=1}^{n_1} \frac{c_s}{t_s^0}} - c_i, \quad i = \overline{1, n_1},$$

$$f_i = \frac{c_i}{t_i^0} \frac{F + \sum_{s=n_1+1}^{n_2} c_s}{\sum_{s=n_1+1}^{n_2} \frac{c_s}{t_s^0}} - c_i, \quad i = \overline{n_1 + 1, n_2}.$$

Proof Conditions (5.44)–(5.46) are directly obtained from Theorem 2. Explicit g_i for $i = \overline{1, n_1}$ and f_i for $i = \overline{n_1 + 1, n_2}$ are available under conditions (5.44) and (5.46) from Zakharov and Krylatov (2014a).

Theorem is proved.

According to Zakharov and Krylatov (2015a), for the network of parallel routes, the optimal condition for n_1 could be expressed explicitly from the program (5.34)–(5.42).

Theorem 4 *The number of green routes n_1^* guarantees a minimal level of emissions on the network of parallel routes in the competitive case if and only if it is the decision of the following optimization program:*

$$\min_{n_1} E(g,f) = \min_{n_1}\Big[\sum_{i=1}^{n_1}\alpha_i\cdot t_i(g_i)\cdot e^{\beta_i\frac{l_i}{t_i(g_i)}} + \sum_{i=n_1+1}^{n_2}\alpha_i\cdot t_i(f_i)\cdot e^{\beta_i\frac{l_i}{t_i(f_i)}}\Big], \tag{5.47}$$

with constraints:

$$G^j > \frac{1}{m+1}\sum_{i=1}^{n_1}\left[\left(\frac{t_{n_1}^0}{t_i^0}\left(1+\frac{h_{n_1}}{c_{n_1}}\right)-1\right)c_i - h_i\right], \tag{5.48}$$

$$F^j > \frac{1}{m+1}\sum_{i=n_1+1}^{n_2}\left[\left(\frac{t_{n_2}^0}{t_i^0}\left(1+\frac{h_{n_2}}{c_{n_2}}\right)-1\right)c_i - h_i\right], \tag{5.49}$$

$$\sum_{i=1}^{n_1}\left[\frac{t_i^0}{m+1}\frac{c_i+h_i}{c_i}+\Psi(G,m)\right]\left[\frac{c_i}{t_i^0}\Psi(G^j,1)-\frac{c_i+h_i}{m+1}\right] \le$$
$$\le \sum_{i=n_1+1}^{n_2}\left[\frac{t_i^0}{m+1}\frac{c_i+h_i}{c_i}+\Phi(F,m)\right]\left[\frac{c_i}{t_i^0}\Phi(F^j,1)-\frac{c_i+h_i}{m+1}\right] \quad \forall j\in M, \tag{5.50}$$

where

$$\Psi(x,y) = \frac{x+\frac{y}{m+1}\sum_{s=1}^{n_1}(c_s+h_s)}{\sum_{s=1}^{n_1}\frac{c_s}{t_s^0}},$$

$$\Phi(x,y) = \frac{x+\frac{y}{m+1}\sum_{s=n_1+1}^{n_2}(c_s+h_s)}{\sum_{s=n_1+1}^{n_2}\frac{c_s}{t_s^0}},$$

and

$$g_i = \left[\frac{c_i}{t_i^0}\sum_{j=1}^{m}\Psi(G^j,1) - m\frac{c_i+h_i}{m+1}\right], \quad i=\overline{1,n_1},$$

Table 5.1 Characteristics of the network of 10 parallel routes

i	1	2	3	4	5	6	7	8	9	10
t_i^0 (min)	6	6.5	7	7.5	8	8.5	9	9.5	10	10.5
c_i (veh/hr)	52	29	18	84	25	24	70	91	57	73
l_i (km)	4	4.3	4.6	5	5.3	5.6	6	6.3	6.6	7

$$f_i = \left[\frac{c_i}{t_i^0} \sum_{j=1}^{m} \Phi(F^j, 1) - m \frac{c_i + h_i}{m+1} \right], \quad i = \overline{n_1 + 1, n_2},$$

when the routes are numbered as follows

$$t_1^0 \left(1 + \frac{h_1}{c_1}\right) \leq \ldots \leq t_{n_1}^0 \left(1 + \frac{h_{n_1}}{c_{n_1}}\right), \tag{5.51}$$

$$t_{n_1+1}^0 \left(1 + \frac{h_{n_1+1}}{c_{n_1+1}}\right) \leq \ldots \leq t_{n_2}^0 \left(1 + \frac{h_{n_2}}{c_{n_2}}\right). \tag{5.52}$$

Proof Conditions (5.48)–(5.52) follow directly from Theorem 4. Explicit g_i for $i = \overline{1, n_1}$ and f_i for $i = \overline{n_1 + 1, n_2}$ are available under conditions (5.48), (5.49), (5.51), and (5.52) from (Zakharov and Krylatov 2014a, 2015a).

Theorem is proved.

Theorems 3 and 4 are quite important because they define the rule for green route allocation that guarantees two conditions: (1) green routes offer less travel time; and (2) a minimal level of emissions is reached. Indeed, consider Figs. 5.1 through 5.4. As an example, consider the network of 10 parallel routes. We assume that Table 5.1 reflects the physical characteristics of this network, and the travel demands are: $G = 200$ green vehicles, $F = 300$ non-green vehicles.

Let us change the number of green routes from one to nine and check out the obtained result. It is quite evident that increasing of number of green routes leads to decreasing the travel time of green vehicles and to increasing the travel time of non-green vehicles Figs. 5.2 and 5.4. At the same time, the lower the number of green routes available, the higher is the speed (the less travel time) of non-green vehicles. However, as it follows from Fig. 5.1, the higher speed non-green vehicles have, the more greenhouse gases they emit. Moreover, it could be verified that four green routes are the best balance between low emissions and an acceptable travel time of non-green vehicles. Actually, the level of emissions for non-green vehicles is nearly uniform when the number of green routes changes from four to nine (Fig. 5.1), while travel time increases monotonically. Simultaneously, the harmful contribution of green vehicles is rather insignificant. Therefore, $n_1 = 4$ would be a quite compromise number for decision maker (Fig. 5.3).

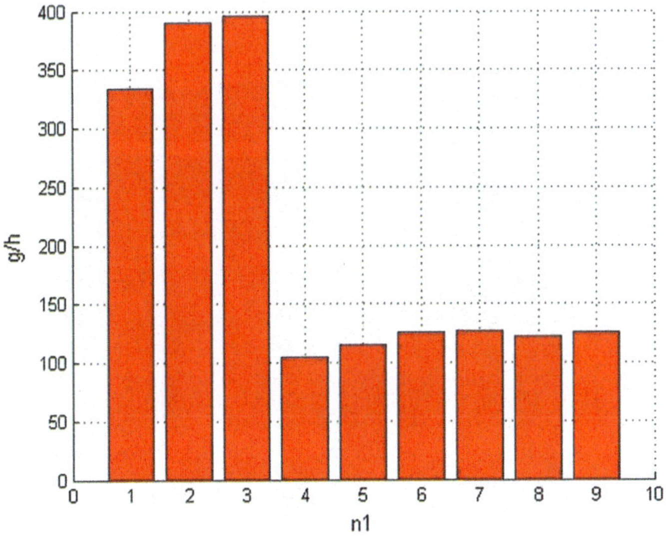

Fig. 5.1 The relationship between travel time and level of emission: total emissions on *non-green* routes

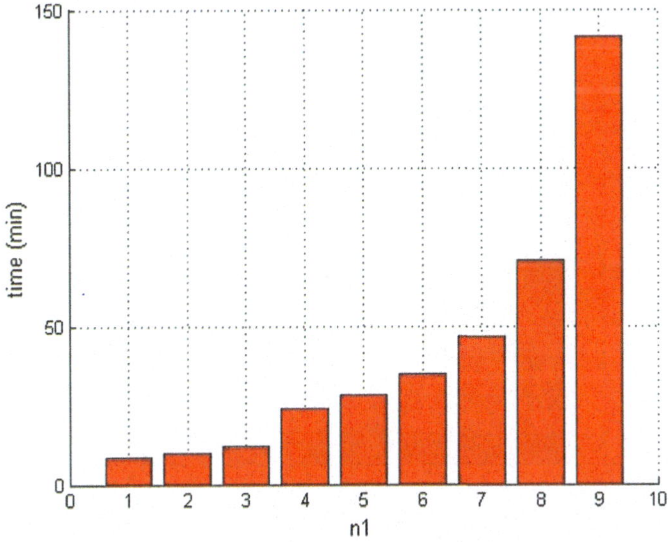

Fig. 5.2 The relationship between travel time and level of emission: travel time through *non-green* routes

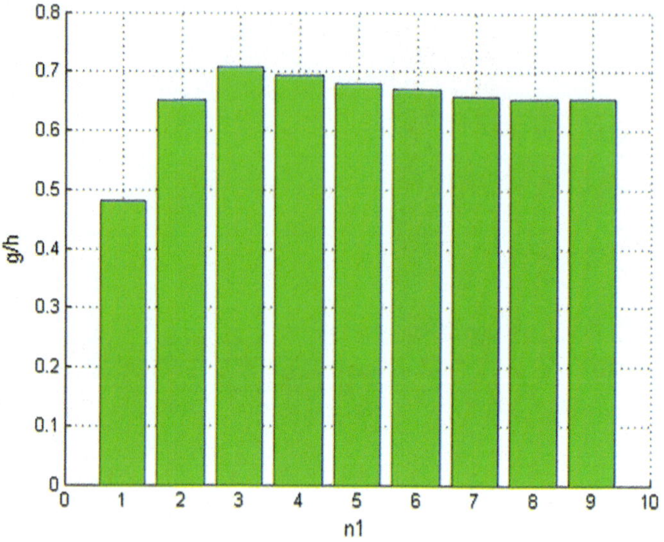

Fig. 5.3 The relationship between travel time and level of emission: total emissions on *green* routes

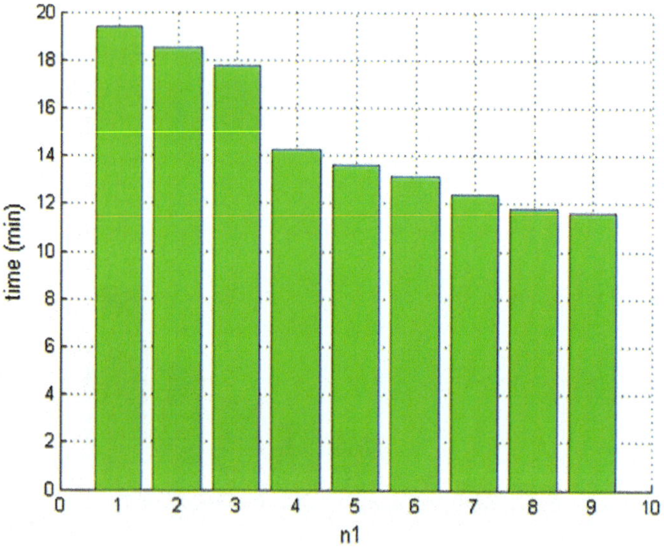

Fig. 5.4 The relationship between travel time and level of emission: travel time through *green* routes

This line of reasoning reflects the path of decision-making that is available due to Theorem 3. An analogical line could be created for Theorem 4.

5.4 Conclusion

Because there is a lack of methodological tools that can be used to support decision makers in decreasing greenhouse gas emission levels (Beltran et al. 2009), this paper investigated the problem of determining the allocation of available green route capacity. The emission function is introduced, and an analysis of green route allocation that guarantees less travel time and a minimal level of emissions is performed. The approach for designing a green transit network that offers green vehicles shorter travel times between given origins and destinations is discussed.

References

Ahn K, Rakha H (2008) The effects of route choice decisions on vehicle energy consumption and emissions. Transp Res Part D Transp Environ 13(3):151–167

Beltran B, Carrese S, Cipriani E, Petrelli M (2009) Transit network design with allocation of green vehicles: a genetic algorithm approach. Transp Res Part C Emerg Technol 17(5):475–483

Ben-Akiva M, De Palma A, Isam K (1991) Dynamic network models and driver information systems. Transp Res Part A Gen 25:251–266

Boriboonsomsin K, Barth M (2009) Impacts of road grade on fuel consumption and carbon dioxide emissions evidenced by use of advanced navigation systems. Transp Res Rec J Transp Res Board: 21–30

Boroujeni BY, Frey HC (2014) Road grade quantification based on global positioning system data obtained from real-world vehicle fuel use and emissions measurements. Atmos Environ 85:179–186

Gaker D, Vautin D, Vij A, Walker JL (2011) The power and value of green in promoting sustainable transport behavior. Environ Res Lett 6(3):034010

Krylatov AY (2016) Network flow assignment as a fixed point problem. J Appl Ind Math 10(2):243–256

Krylatov A, Zakharov V (2016) Competitive traffic assignment in a green transit network. Int Game Theory Rev

Lin J, Ge YE (2006) Impacts of traffic heterogeneity on roadside air pollution concentration. Transp Res Part D Transp Environ 11(2):166–170

Mahmassani HS (1990) Dynamic models of commuter behavior: experimental investigation and application to the analysis of planned traffic disruptions. Transp Res Part A Gen 24(6):465–484

Sheffi Y (1985) Urban transportation networks: equilibrium analysis with mathematical programming methods

US Bureau of Public Roads (1964) Traffic assignment manual

Wyatt DW, Li H, Tate JE (2014) The impact of road grade on carbon dioxide (CO 2) emission of a passenger vehicle in real-world driving. Transp Res Part D Transp Environ 32:160–170

Yin Y, Lawphongpanich S (2006) Internalizing emission externality on road networks. Transp Res Part D Transp Environ 11(4):292–301

Zakharov V, Krylatov A (2014a) Equilibrium assignments in competitive and cooperative traffic flow routing. In: Working conference on virtual enterprises. Springer, Heidelberg, pp 641–648

Zakharov V, Krylatov A (2014b) OD-matrix estimation based on plate scanning. In: 2014 international conference on computer technologies in physical and engineering applications (ICCTPEA)
Zakharov V, Krylatov A (2015a) Competitive green-vehicle assignment on a transportation network. Game Theor Models Math Ecol: 360–396
Zakharov V, Krylatov A (2015b) Transit network design for green vehicles routing. Springer International Publishing, pp 449–458
Zhang K, Frey HC (2006) Road grade estimation for on-road vehicle emissions modeling using light detection and ranging data. J Air Waste Manag Assoc 56(6):777–788
Zhang Y, Lv J, Ying Q (2010) Traffic assignment considering air quality. Transp Res Part D Transp Environ 15(8):497–502

Chapter 6
Why to Climb If One Can Jump: A Hill Jumping Algorithm for the Vehicle Routing Problem with Time Windows

David Mester, Olli Bräysy and Wout Dullaert

Abstract The most common approaches to solve the variants of the well-known vehicle routing problem are based on metaheuristic hill-climbing search. The deficiency of these methods is slow local search based hill climbing that often is restricted to limited local neighborhood. In this paper we suggest a novel new two-phase metaheuristic that escapes the local minima with jumps of varying size, instead of step by step local hill climbing. The initial solution is first generated with a powerful ejection pool heuristic. The key idea of the improvement phase is to combine large neighborhood search with standard guided local search metaheuristic in a novel way, allowing improved search diversification and escape from local minima in more efficient way through jumps. The algorithm has been tested on the standard Gehring and Homberger benchmarks for the vehicle routing problem with time windows and the results indicate very competitive performance. We found 12 new and 43 matched best-known solutions and the best overall results for all problem sizes at comparable computation times.

6.1 Introduction

The vehicle routing problems (VRP) are at the heart of the less than truckload distribution. There is a growing importance of managing them effectively due to increased competition and growing volumes, e.g. due to e-commerce. Furthermore, current

D. Mester (✉)
Institute of Evolution, Mathematical and Population Genetics Laboratory,
University of Haifa, 31905 Haifa, Israel
e-mail: dmester@research.haifa.ac.il

O. Bräysy
Myopt Consulting Oy, Sepänkatu 4, 40720 Jyväskylä, Finland

W. Dullaert
Faculty of Economics and Business Administration,
VU University Amsterdam, De Boelelaan 1105, 1081 Amsterdam,
HV, The Netherlands
e-mail: wout.dullaert@vu.nl

P. Diez et al. (eds.), *Computational Methods and Models for Transport*,
Computational Methods in Applied Sciences 45,
DOI 10.1007/978-3-319-54490-8_6

concerns over global warming, resource depletion, and the social impact of traffic congestion and pollution are driving companies, governments, and researchers to improve the efficiency of logistics and distribution operations.

Given the NP-hard nature and complexity of the VRP (Lenstra and Kan 1981), to practical problems, heuristics are the only viable solution approach. Traditionally, VRPs have been solved with 2-phase metaheuristics that are based on local search type hill-climbing. The most common metaheuristics include tabu search, genetic algorithm and variants of simulated annealing (Bräysy and Gendreau 2005b). Correspondingly, the most common local searches consist of simple k-opt and exchange neighborhoods (Bräysy and Gendreau 2005a). The main deficiency of these approaches is that the local search is time consuming and due to various tight constraints, it can only explore very limited search space in the vicinity of the existing base solution, even if some worsening to objective function value is allowed. As the key idea of hill climbing is to change the solution structure to allow wider search space exploration, we suggest here a novel and more straightforward strategy where instead of step-by-step hill climbing, we attempt larger single-step jumps of varying size to the current solution. This is followed by standard local search to locally optimize the new solution. The same strategy is also used in the suggested initial solution procedure that combines several insertions into chain to allow better exploration of the search space while minimizing the number of routes. Finally, a systematic memory structure is used to further guide the search process.

The following solution concepts have been identified as being most successful (Vidal et al. 2013).

Evolution strategies: Mester and Bräysy (2005), Labadi et al. (2008), Repoussis et al. (2009), Nagata et al. (2010), Vidal et al. (2013); *Solution recombinations:* Labadi et al. (2008), Repoussis et al. (2009), Nagata et al. (2010); *Ruin and recreate*: Pisinger and Ropke (2007), Repoussis et al. (2009), Prescott-Gagnon et al. (2009); *Ejection chains:* Nagata and Bräysy (2009), Nagata et al. (2010); *Guidance and memories:* Mester and Bräysy (2005), Repoussis et al. (2009).

For comprehensive survey, we refer to Bräysy and Gendreau (2005a, b). A closer analysis reveals that the suggested new metaheuristic is based on combining ideas from all five above listed most successful domains. First, it applies the ejection chain methodology of Nagata and Bräysy (2009) to generate initial solutions with a low number of routes. To minimize the distance objective, the ruin and recreate principle Shaw (1997) is combined with the guided local search metaheuristic (Voudouris and Tsang 1999), utilizing memory based guidance. Finally, the solution acceptance process embedded in the ruin and recreate involve principles from evolution strategies and solution recombinations.

The new solution method will be tested with the VRP with time windows (VRPTW) as it is one of the most commonly studied and important VRP variants. The overall objective of the VRPTW is to find a minimum set of vehicle routes with identical vehicle capacity, to service a group of customers with a given capacity demand, at minimum total travel distance. The services of the customers have to be started within the customer available time window, the vehicle is allowed to arrive before the opening of the time window, and wait at no cost until the customer is available. However, the vehicle is not allowed to arrive after the ending of the customer available time window. The primary objective is to minimize the number of vehicles required to serve all customers, and the secondary objective is to minimize the total distance travelled by the vehicles. This class of problems has many applications in practical logistics related activities, such as container truck routing, delivery service scheduling and many logistics system problems.

The suggested method has been tested on the well-known benchmark problems of Gehring and Homberger (1999). Moreover, a sensitivity analysis is presented. The remainder of this paper is organized as follows. In the next section we detail the new metaheuristic and its components. Section 6.3 presents the sensitivity analysis and overall computational results and the conclusions are drawn in Sect. 6.4.

6.2 The New Metaheuristic

The suggested new metaheuristic consists of two main phases. The initial solution is generated with the ejection pool heuristic of Nagata and Bräysy (2009). The obtained initial solution is then further improved in terms of distance by the new two-phase metaheuristic that oscillates between a local search and large neighborhood search phases.

6.2.1 The Initial Solution Procedure

The suggested route elimination algorithm starts with an initial solution where each customer is served individually by a separate route. Then an attempt is made to reduce the number of routes in the solution one by one in random order until the total computation time reaches a given limit. Each time all the customers in the route to be eliminated are used to initialize the ejection pool (EP). The main idea of the ejection pool is to always hold the set of unserved (ejected) customers, currently missing from the solution. In each iteration of the algorithm, attempts are made to insert customers from the pool back to the solution. At first only straightforward insertions between two consecutive nodes are considered and all feasible insertions are identified. Here the actual implemented insertion is selected randomly among the feasible options. If no feasible direct insertions can be found, a more sophisticated insertion procedure is applied in which an infeasible insertion is temporarily allowed, followed by a series of local search moves to restore the feasibility of the partial solution without removing any customer from the solution.

If the sophisticated insertions also fail, we allow removals (ejections) of customers from the current solution up to a user-defined maximum limit, in order to insert a customer from the pool and to limit the search. The ejected customer is chosen based on the search history so that a customer or a customer pair is selected that is easiest to reinsert based on previous reinsertion failure counts. The customers are chosen for insertion from the ejection pool according to the last in, first out (LIFO) strategy. Moreover, after each customer removal, the search is diversified by applying random feasibility-preserving local search moves to the entire solution. For more details, we refer to Nagata and Bräysy (2009).

6.2.2 Metaheuristic Improvement

The basic principle of the applied large neighborhood scheme is quite simple. A set of customers is removed from the current solution according to a chosen removal strategy, after which an attempt is made to reinsert the removed customers back to the solution applying a standard cheapest insertion heuristic with modified insertion criterion of Osman (1993). The key idea is to select an insertion where

$$\cos t_{\text{arcs}}^{\text{old}} - \alpha \cdot \cos t_{\text{arcs}}^{\text{new}} \quad (6.1)$$

is maximized. Here α is a user-defined parameter to which different values are attempted during the search. The suggested new method involves four important innovations compared to the standard approaches. First, each time after all points have been inserted back to the solution, the three local searches are executed up to local minimum. Second, the removal, reinsertion and local search steps are repeated with both removal strategies and the better of the obtained two solutions replaces the current solution, regardless of whether the new solution is better than the current one. Each time a new solution is obtained, its objective value is compared against the best found solution and it is updated when necessary. In the end of the search, the best found solution is returned. The third innovation is that both in the insertion and local search evaluations, the GLS-based objective function is applied. The fourth issue is a restart mechanism. If for a given number of iterations R, no improvement has been found, the search is restarted from the best found solution. At the same time the previous value of R is increased by 5%, the arc penalty history is cleared and the value of the penalty parameter λ is decreased by 10%.

Both removal strategies are based on removing a randomly selected set and number Z of points from the current solution. The value of Z is defined according to equation (6.2).

$$Z = (0.25 + 0.75r) \cdot n \quad (6.2)$$

Here r is a random number between 0 and 1 and n is the number of customers. If $Z < 0.5n$, Z points are randomly removed from the solution. Otherwise, the removals are restricted to closest routes around the currently penalized arc, serving Z customers. The search is terminated if no more improvement has been found for a given number of iterations. Here the term iteration refers to repeating the LNS and local search steps with both removal strategies.

The distance improvement phase combines a set of standard local search heuristics and a large neighborhood search heuristic (Shaw 1997) with the Guided Local Search (GLS) metaheuristic guiding mechanism.

Because long arcs should not occur our GLS operates by penalizing particular solution features (long arcs) it considers should not occur in a near-optimal solution. The penalization is weighted by the number of times the arc has already been penalized. The more often an arc appears and is penalized, the less likely it is to be penalized further. The penalties are initially set to 0 and are updated during the search each time a local minimum is reached. More precisely, given an objective function g that maps every candidate solution s to a numerical value, GLS replaces g with a new objective function h

$$h(s) = g(s) + \lambda \cdot (p_i I_i(s)) \tag{6.3}$$

where s is a candidate solution, λ is a parameter controlling the impact of penalties, p_i is the penalty counter for feature i, which holds the number of times i has been penalized, and I_i is a indicator function, which tests whether s exhibits feature i:

$$I_i(s) = 1 \quad \text{if } s \text{ exhibits feature } i \text{ : and 0 otherwise.}$$

For more details on the applied GLS strategy, we refer to Mester and Bräysy (2005).

The set of applied local search heuristics consist of relocate Savelsbergh (1992), 1-interchange (Osman 1993) and 2-opt* (Potvin and Rousseau 1995). These operators are applied cyclically with the best-accept strategy such that each operator is applied only once at a time, and then the neighborhood is changed.

6.3 Experimental Results

As the focus of our paper is on large-scale problems, the algorithm has been tested on the 300 extended and well-studied VRPTW benchmarks of Gehring and Homberger (1999). The Gehring and Homberger test instances consist of 200, 400, 600, 800 and 1000 customers. For each problem size, the 60 problems are divided into $R1$, $C1$, $RC1$, $R2$, $C2$ and $RC2$ groups. Travel times separating two customers corresponds to their relative Euclidean distance. The problems vary in fleet size, vehicle capacity, spatial and temporal distribution of customers, time window density and width and customer service times. In problems belonging to the C group, the customers are clustered geographically or based on their time windows. In the R group the customers are uniformly distributed, and the RC group combines clustered and randomly distributed customers. Problem sets $R1$, $C1$ and $RC1$ have a narrow scheduling horizon. Hence, only a few customers can be serviced by the same vehicle. Conversely, problem sets $R2$, $C2$ and $RC2$ have a larger scheduling horizon, and more customers can be served by the same vehicle.

We used a hierarchic objective function in our experiments, where the primary objective is to minimize the number of routes. For the same number of routes, the secondary objective is the minimization of the total traveled distance. The proposed algorithm has been implemented in Visual Basic 6.0 and it includes a graphical user interface that both have a considerable negative impact on the computation time compared to an efficient C++ implementation. The experimental tests consist only of one simulation run for each problem with a Intel 2.66 GHz computer.

The start values of the two parameters are set to $R = 5000$ and $\lambda = 0.2$. We also set a limit that the value of λ cannot be decreased below 0.0005. The time limit for the initial solution phase was set to 5 min. These parameter values were set by running comprehensive tests over all problem instances and choosing the value that gives the best overall output.

Figure 6.1 illustrates the sensitivity of the results on the average over all test problems to the size of the removal set, Z. Here n refers to the number of customers in the problem. Here blue curve represents the solution quality and the red curve the CPU time. The values of the curves show the percentual differences with the lowest value obtained, but so that the red curve values are scaled to range [0, 1] for better visualization. The figure shows that one obtains the better solutions the more points are considered for removal, but the differences are not significant, remaining within 0.7%. At the same time the CPU time grows by over 30% compared to the smallest value, but gets down by about 12% in the end. We believe this is due to more efficient overall search efficiency.

The average sensitivity of the results to different values of α is illustrated in Fig. 6.2 over all test problems via relative percentual differences in the same way as in Fig. 6.1. Based on the figure, the algorithm is very sensitive to the different lambda values and even to small differences. At worst, the impact can be close to 1.6%. Value 0.2 appears to be the best from solution quality viewpoint, requiring at the same time most CPU time. Also in general the λ values providing better solution quality seem to require more CPU time.

The significance of the different solution components is illustrated in Table 6.1. The table clearly shows that LNS is the most important component and the results worsen by over 2% on the average if LNS component is omitted. At the same time,

Fig. 6.1 The impact of the neighborhood size to solution quality and CPU time

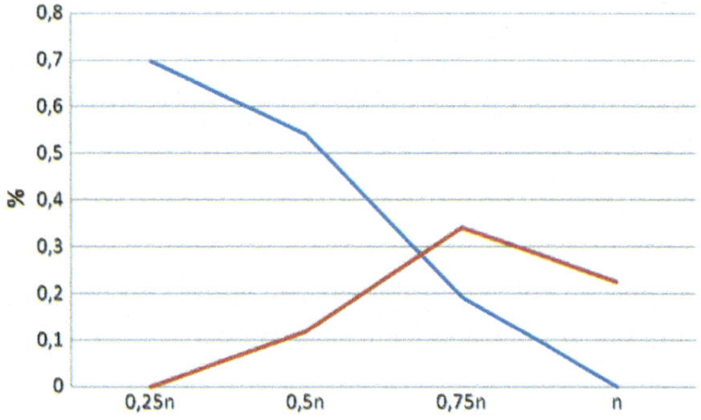

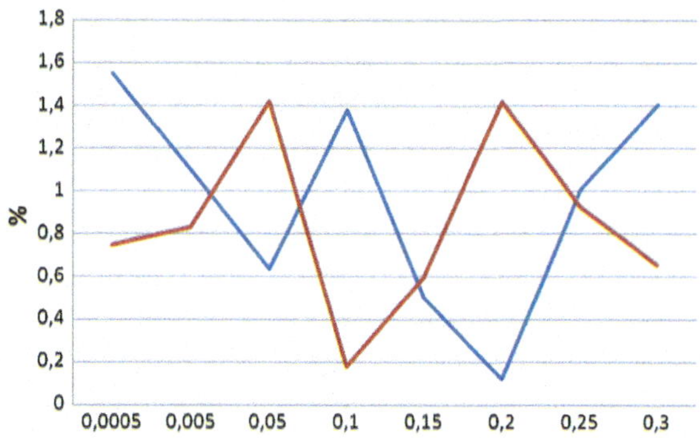

Fig. 6.2 The impact of different lambda values to solution quality and CPU time

Table 6.1 The significance of the solution components

Missing component	No Multi-start (%)	No GLS (%)	No LNS (%)
Solution worsening	0.92	1.17	2.08
CPU reduction	11	14	33

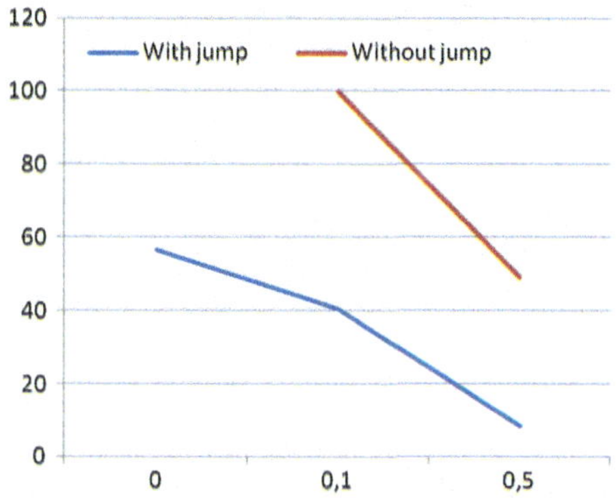

Fig. 6.3 Importance of the jump mechanism

omitting the use of LNS also reduces the CPU time by over 30%. Based on the table, the multi-start and GLS are also important parts of the algorithm, with a minor impact to the CPU time. On the other hand, none of the components has a drastically worsening impact.

Figure 6.3 provides a more detailed analysis on the importance of the suggested new jump-search strategy. This figure shows the relative CPU time differences with respect to largest CPU time observed (100%) to obtain the solutions that equal the best-known solution or are within 0.1% or 0.5% from the best-known solution. The experiment was run on instances to which our algorithm was able to find the best-known solution.

Table 6.2 The detailed results

Size	R1	R2	C1	C2	RC1	RC2
200	*18.2*	*4.0*	*18.8*	*6.0*	*18.0*	*4.3*
	3611.75	2939.03	2716.82	1773.32	3184.83	2546.32
400	*36.3*	*8.0*	*37.6*	*11.6*	*36.0*	*8.5*
	8425.21	6187.49	7182.02	3946.17	7906.40	5324.50
600	*54.5*	*11.0*	*57.5*	*17.4*	*54.4*	*11.4*
	18100.30	12487.10	14180.16	7741.02	16157.69	10953.61
800	*72.8*	*15.0*	*75.0*	*23.3*	*72.0*	*15.4*
	31466.37	20289.65	25636.01	11658.30	29283.10	16925.18
1000	*91.9*	*19.0*	*93.9*	*28.8*	*90.0*	*18.2*
	47683.64	29603.05	41449.55	17253.88	44669.64	24803.10

According to Fig. 6.3, it takes approximately twice more time to obtain a solution within 0.5% limit from the best-known solution if LNS-based jumping mechanism is not applied. The relative difference is even larger in case of 0.1% limit and without the LNS-jumping, the algorithm was not able to find the best-known solutions even if the computing times was increased up to 10 h. This clearly shows the importance and efficiency of the new jump strategy.

Table 6.2 reports the problem type and size specific results to the tested benchmark set. For each five problem size, we report the average number of vehicles and average total distance to each six problem classes to enable a detailed comparison of the results with the previous methodologies.

In Table 6.3 we compare the overall results against the three recent and best performing methods. The comparison is based on the cumulative number of vehicles (CNV) and cumulative distance (CTD) over all problems of given size and the average CPU time in minutes. If multiple runs or processors are applied, the CPU times in the table are multiplied by the corresponding number. According to Table 6.3, the suggested new method outperforms the previous methods in terms of CNV for all problem sizes that is the primary optimization objective. It also outperforms the previous methods in terms of CTD for the 200-customer problems. For other problems, the CTD values of our method are higher. One must however remember that often lower number of vehicles is obtained at the cost of longer traveling, so the results with different CNV values are not directly comparable.

As for the CPU time viewpoint, our approach appears to be competitive and comparable with the other methods, even considering the differences in computers. We found also 12 new best-known solutions, listed in Table 6.4.

Table 6.3 Comparison among the best methodologies

Author		200	400	600	800	1000
Repoussis et al. (2009)	CNV	694	1381	2066	2739	3428
	CTD	169163	395936	816326	1424321	2144830
	CPU	90	180	270	360	450
Nagata and Bräysy (2009)	CNV	694	1381	2067	2739	3424
	CTD	168143	388548	789420	1352478	2040661
	CPU	4.7	34	80.4	126.8	186.4
Vidal et al. (2013)	CNV	694	1381	2068	2739	3420
	CTD	168092	388013	786373	1334963	2036700
	CPU	43	170.5	497	1075	1745
Mester and Bräysy (2005)	CNV	**693**	**1380**	**2062**	**2735**	**3418**
	CTD	**167721**	389717	796199	1352586	2054629
	CPU	9.4	54.1	96	125	389.9

Table 6.4 The new best-known solutions

Problem	Vehicles	Distance	Problem	Vehicles	Distance
R122	18	4039.86	R1103	91	45176.38
R146	36	8373.41	R1104	91	42787.19
RC148	36	7760.23	R1105	91	51830.36
C249	12	3864.67	R1107	91	44435.50
R168	54	15716.15	R1108	91	42484.37
C266	18	7470.36	R11010	91	48294.71

6.4 Conclusions

We have presented a new powerful hybrid metaheuristic that successfully combines local search with ruin and recreate principle and memory based guidance mechanism. The suggested method is simple to implement and contains several sophisticated features. It has been tested on the well-studied benchmark problems and the obtained results outperform the previous best methods. Moreover, a significant number of new best-known solutions are obtained. The future work involves further improvement of the performance e.g. via parallel computing and applying the algorithm to other related problems.

Acknowledgements This research was partially supported by the TULOPT TEKES program and Procomp Solutions Ltd.

References

Bräysy O, Gendreau M (2005a) Vehicle routing problem with time windows, Part I: route construction and local search algorithms. Transp Sci 39(1):104–118

Bräysy O, Gendreau M (2005b) Vehicle routing problem with time windows. Part II: Metaheuristics. Transp Sci 39(1):119–139

Elhamifar E, Vidal R (2013) Sparse subspace clustering: algorithm, theory, and applications. IEEE Trans Pattern Anal Mach Intell 35(11):2765–2781

Gehring H, Homberger J (1999) A parallel hybrid evolutionary metaheuristic for the vehicle routing problem with time windows. In: Proceedings of EUROGEN99, vol 2, pp 57–64, Citeseer

Golden BL, Raghavan S, Wasil EA (2008) The vehicle routing problem: latest advances and new challenges, vol 43. Springer Science & Business Media

Labadi N, Prins C, Reghioui M (2008) A memetic algorithm for the vehicle routing problem with time windows. RAIRO-Oper Res 42(3):415–431

Lenstra JK, Kan AHG (1981) Complexity of vehicle routing and scheduling problems. Networks 11(2):221–227

Mester D (1999) A parallel dichotomy algorithm for vehicle routing problem with time windows. Technical report, Working paper, Minerva Optimization Center, Technion, Israel

Mester D, Bräysy O (2005) Active guided evolution strategies for large-scale vehicle routing problems with time windows. Comput Oper Res 32(6):1593–1614

Nagata Y, Bräysy O (2009) A powerful route minimization heuristic for the vehicle routing problem with time windows. Oper Res Lett 37(5):333–338

Nagata Y, Bräysy O, Dullaert W (2010) A penalty-based edge assembly memetic algorithm for the vehicle routing problem with time windows. Comput Oper Res 37(4):724–737

Osman IH (1993) Metastrategy simulated annealing and tabu search algorithms for the vehicle routing problem. Ann Oper Res 41(4):421–451

Pisinger D, Ropke S (2007) A general heuristic for vehicle routing problems. Comput Oper Res 34(8):2403–2435

Potvin J-Y, Rousseau J-M (1995) An exchange heuristic for routeing problems with time windows. J Oper Res Soc 46(12):1433–1446

Prescott-Gagnon E, Desaulniers G, Rousseau L-M (2009) A branch-and-price-based large neighborhood search algorithm for the vehicle routing problem with time windows. Networks 54(4):190–204

Repoussis PP, Tarantilis CD, Ioannou G (2009) Arc-guided evolutionary algorithm for the vehicle routing problem with time windows. IEEE Trans Evol Comput 13(3):624–647

Savelsbergh MWP (1992) The vehicle routing problem with time windows: minimizing route duration. ORSA J Comput 4(2):146–154

Shaw P (1997) A new local search algorithm providing high quality solutions to vehicle routing problems. Dept of Computer Science, University of Strathclyde, Glasgow, Scotland, UK, APES Group

Vidal T, Crainic TG, Gendreau M, Prins C (2013) A hybrid genetic algorithm with adaptive diversity management for a large class of vehicle routing problems with time-windows. Comput Oper Res 40(1):475–489

Voudouris C, Tsang E (1999) Guided local search and its application to the traveling salesman problem. Eur J Oper Res 113(2):469–499

Chapter 7
Clustering Driving Destinations Using a Modified DBSCAN Algorithm with Locally-Defined Map-Based Thresholds

Ghazaleh Panahandeh and Niklas Åkerblom

Abstract The aim of this paper is to propose a method to cluster GPS data corresponding to driving destinations. A new DBSCAN-based algorithm is proposed to group stationary GPS traces, collected prior to end of trips, into destination clusters. While the original DBSCAN clustering algorithm uses a global threshold as a closeness measure in data space, we develop a method to set local thresholds values for data points; this is important because the GPS data proximity strongly depends on the density of the street grid around each point. Specifically, the spread of GPS coordinates in parking lots can vary substantially between narrow (personal parking lot) and wide (parking lot of a shopping mall) depending on the destinations. To characterize the parking lot diversities at each destination, we introduce the concept of using a local threshold value for each data point. The local threshold values are inferred from road graph density using a map database. Moreover, we propose a mutual reachability constraint to preserve the insensitivity of DBSCAN with respect to the ordering of the points. The performance of the proposed clustering algorithm has been evaluated extensively using trips of actual cars in Sweden, and some of the results are presented here.

7.1 Introduction

Nowadays, huge amounts of location-based data are being shared through the cellular networks with GPS receivers in car navigation systems. The availability of such data opens up new research areas in pattern analysis and data mining. Analyzing individual driving/mobility-patterns from logged GPS data have found a wide range of applications, such as path or destination prediction, real time traffic volume estimation, city planning, energy consumption optimization, etc. In these systems, predictive models are constructed mainly based on statistical properties of data given that it

G. Panahandeh (✉) · N. Åkerblom
Volvo Car Corporation, 40531 Gothenburg, Sweden
e-mail: gpanahan@volvocars.com

N. Åkerblom
e-mail: nakerblo@volvocars.com

P. Diez et al. (eds.), *Computational Methods and Models for Transport*,
Computational Methods in Applied Sciences 45,
DOI 10.1007/978-3-319-54490-8_7

follows some regularity patterns. The patterns can be inferred by analyzing driving history (routes from origins to destinations). Despite large similarities in trajectories when driving from an origin to a destination, diversity of parking locations can vary depending on vicinity of parking lots for different destinations. Therefore, grouping end-of-trip locations is an essential and primary step for driving data analysis, especially when constructing destination-dependent probabilistic models.

The focus of our study is to develop a destination clustering algorithm by analyzing the history of end-of-trip logged GPS data for individual drivers. That is, we are aiming to label the parking locations into physically meaningful clusters (destinations/origins). We propose a modified DBSCAN algorithm by introducing a new concept on using local density thresholds. The local threshold for each point is adapted according to the density of the street grid around that point, thereby capturing inherent differences between driver destinations in disparate areas, e.g. urban, residential or commercial zones.

Related works in this field attempt to cluster places into regions of interest for groups of people (Palma et al. 2008), i.e. identifying clusters with high densities. However in our application, the clustering is performed for individuals and it is independent of the data density at each destination. Other clustering methods use trajectory-similarity for grouping nearby locations (Rinzivillo et al. 2008; Giannotti et al. 2007). However, these methods have high computational complexity and result in clusters that are origin-dependent. It should also be mentioned that, the problem in hand differs from those clustering methods that are introduced for destinations clustering when using a handheld device. In these cases, a destination can be defined once the user enters an indoor location and stay there for at least a certain period of time. However, when collecting data in cars equipped with GPS devices, locations corresponding to ignition-off event varies depending on parking spots for each destination. Hence, diversity of logged GPS data corresponding to a destination could be very different depending on density of road graphs in each area.

7.2 Proposed Clustering Algorithm

For our clustering application, data is recorded from cars equipped with portable GPS devices that are connected to a portal server. Data attributes are geographical information corresponding to stationary GPS traces collected prior to end of trip. A trip starts from an ignition-on event at a location corresponding to an origin cluster and ends by an ignition-off event at a location corresponding to a destination cluster. Assuming connectivity properties for each pair of origin-destination, we hereafter use the term *destination* when referring to both the origin and the destination.

A cluster is a group of data that share a set of similar properties. In our applications, data attributes are numeric (GPS locations) where the similarity between two data object can be judged based on a distance measure. Moreover in this problem, the number of clusters are not known in advance and data points are sequentially increasing over time. Given these requirements and the data type properties, we choose to use DBSCAN-based clustering algorithms because of their ability in discovering clusters with arbitrary shapes and processing large databases efficiently. The original DBSCAN (Ester et al. 1996) algorithm and its variations (OPTICS (Ankerst et al. 1999), LDBSCAN (Duan et al. 2007), PDBSCAN (Kisilevich et al. 2010)) have been widely used for clustering spatial data. In these algorithms, the similarity measures are either defined globally in data space or determined locally based on density of data at different regions. In our application, finding appropriate threshold value as a closeness measure can be very challenging depending on the location of the parking lot, which is independent of density of previously collected data at each region. Therefore, the current density-based algorithms do not provide a complete solution in our application considering the region-based clusters specifications.

In this paper, we propose a modified DBSCAN clustering approach in which the local similarity measures are adapted according to the density of the road grid around each data point. Hence, the defined local similarities are independent of points density at each cluster. In the following, we briefly describe the original DBSCAN algorithm (Ester et al. 1996) and then present our proposed algorithm.

In DBSCAN, clusters are formed by all points that are density-reachable from each other, i.e. connected through an unbroken chain of directly density-reachable points. In turn, a point p is directly density-reachable from a point q, if q has a minimum number of points (defined as $minPts$) including p within a distance threshold of ϵ. The $minPts$ and ϵ values are given as global parameters into DBSCAN. The set of all points that are directly density-reachable from q is called the ϵ-neighborhood of q. All points which are not density-reachable from any other point are classified as noise.

Figures 7.1 and 7.2 show the results of DBSCAN clustering in a subset of GPS coordinates for one user centered in the town of Alingsås (vicinity of 5 km^2). The $minPts$ value was set to 3 for both of these examples. To depict the deficiency of DBSCAN with respect to our clustering application, we choose two distinct ϵ values for each experiment. Figure 7.1 illustrates the resulting clusters with $\epsilon = 100$ m, where noises (depicted in black) and six distinct dense clusters (depicted with other colors) can be seen. The destination clusters correspond well to the self-identified activities of the user, except in the lower right corner, where two parking lots belonging to the same building are clustered as separate destinations. In an effort to correct that misclassification, the ϵ parameter is increased to 200 m, see Fig. 7.2. This results in merging the lower right clusters correctly, while it also has the side effect of clustering many of the noise points (corresponding to distinct destination) in town as one large cluster. As can be observed, resulting destinations are highly dependent on properties of parking lots areas (density of the street grid around these point), which cannot be captured using only one global ϵ value.

To adapt the ϵ value in DBSCAN according to density of the street grid around each GPS location, we propose a two-step solution. The first step is to modify the DBSCAN algorithm by defining an individual density threshold for each data point. The second step is to identify those thresholds using knowledge of the problem domain. To preserve symmetricity of the original DBSCAN algorithm when defining the individual density threshold for each data point, it is required to redefine the notion of ϵ-neighborhood from DBSCAN according to $N_\epsilon(p) = \{q \in D | dist(p, q) < \epsilon_p,\ \ dist(p, q) < \epsilon_q\}$. That is p and q should be reachable from both directions to be considered members of the same neighborhood. It is worth mentioning that the run time efficiency of the DBSCAN is retained by performing the region query with ϵ_p and subsequently excluding results that do not fulfill the condition $dist(p, q) < \epsilon_q$.

As has been previously stated, in densely-built areas, e.g. in town and city centers, the result of a high ϵ value could be that the entire area is clustered together, while the opposite applies for low values of ϵ in less dense areas, i.e., they should be clustered together. This observation implies that there is an inverse relationship between the density of the area in which a point p is situated and the density threshold ϵ_p. As a density measure, we use a map database to retrieve the number of road links within a constant *radius* of each point in the data set. These density values are then used in a linear interpolation between a lower ϵ-bound for dense areas and an upper ϵ-bound for less dense areas. The ϵ-bounds and their corresponding (two)-breakpoints for the density values are set intuitively according to the data. The density values below and above the breakpoints are set to the lower and upper ϵ-bounds respectively, resulting in the application of the original DBSCAN algorithm on those points. Figure 7.3 shows resulting clusters when applying our proposed algorithm, referred to as the map-based local DBSCAN (ML-DBSCAN), to the same dataset. As can be seen, the previous issues with low and high ϵ values in Figs. 7.1 and 7.2 are resolved. For this

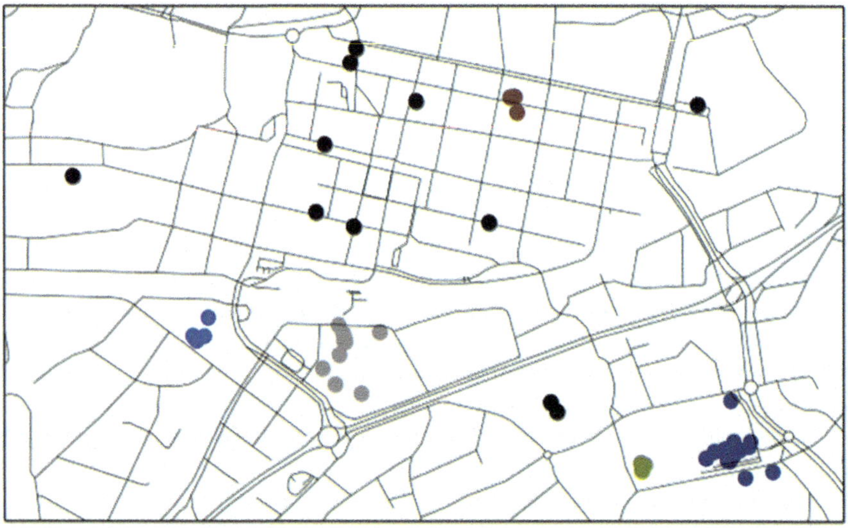

Fig. 7.1 DBSCAN $\epsilon = 100$ m. Points with similar color are categorized into one cluster where noise points are depicted in *black*. The points in the *lower right corner*, which belong to one parking lot, are detected as two clusters

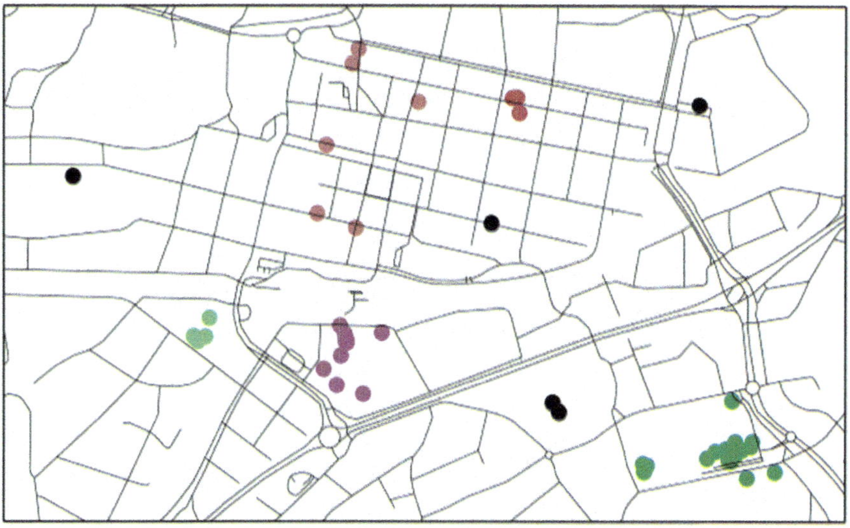

Fig. 7.2 DBSCAN $\epsilon = 200$ m. Points with similar color are categorized into one cluster where noise points are depicted in *black*. By increasing the ϵ value from 100 to 200, compared to Fig. 7.1, the points in the parking lot are clustered together, while many of the noise points (corresponding to distinct destination) in the city center are merged into one large cluster

implementation, the first and second breakpoints are set to 100 and 200, respectively, using the radius of 500 m, and the lower and upper bounds are set to 100 m and 200 m, respectively.

We have extensively evaluated the performance of our proposed clustering algorithm. Herein, some of the results are reported for five car users with 514, 366, 847, 599, and 754 detected parking events recorded during 4, 6, 6, 6, and 6 months, respectively. Table 7.1 shows the clustering results for three different radius settings, where interpolation parameters, including first and second breakpoints (1-BP, 2-BP), and upper and lower bounds are intuitively set to the given values. The experiments show that similar clustering results are obtained for all three radius values. These results were also visually confirmed on the map for individual users. To further study the sensitivity of the proposed algorithm to the parameter settings, the interpolation parameters (first and second breakpoints, and upper and lower bounds) were swept 10% around the values given in Table 7.1 at each radius, resulting to 81 different settings. The experiments show that the proposed algorithm leads to stable and visually-meaningful results for a wide range of parameter values, and no major change in the clustering results were observed.

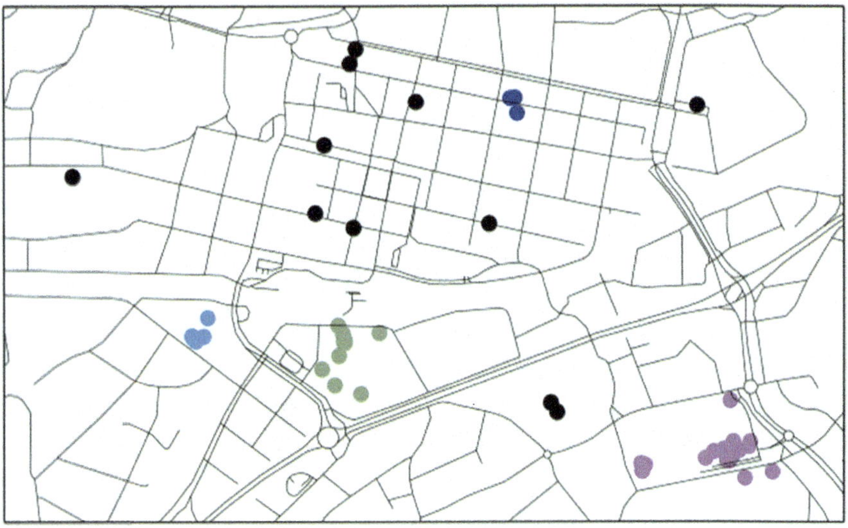

Fig. 7.3 The resulting clusters from our proposed algorithm, ML-DBSCAN. Points with similar color are categorized into one cluster where noise points are depicted in *black*. As can be seen, the previous issues with low and high ϵ values in Figs. 7.1 and 7.2 are resolved

Table 7.1 Resulting clusters with different parameter settings for interpolation parameters, including first and second breakpoints (1-BP, 2-BP), upper and lower bounds

Radius	1-BP	2-BP	Upper bound	Lower bound	Number of clusters				
					User 1	User 2	User 3	User 4	User 5
750	150	300	100	100	31	22	18	27	30
500	100	200	200	100	31	21	17	27	31
250	50	100	200	100	31	21	17	27	30

7.3 Conclusion and Future Work

In this paper, we propose a modified DBSCAN algorithm in which local density thresholds are adapted according to street grid-density around each data point. The results indicate the accuracy and reliability of our proposed map density-based clustering method compared to the original DBSCAN algorithm, specifically where data points are distributed in areas with different densities in the road grid. As a future work, we are investigating adaptive solutions in which street grid densities can be dynamically assigned to each point using the map database. For this purpose, we are analyzing road graph densities and statistical properties of points at different region.

References

Ankerst M, Breunig MM, Kriegel HP, Sander J (1999) OPTICS: ordering points to identify the clustering structure. ACM Sigmod Record 28:49–60

Duan L, Xu L, Guo F, Lee J, Yan B (2007) A local-density based spatial clustering algorithm with noise. Inf Syst 32(7):978–986

Ester M, Kriegel H-P, Sander J, Xu, X et al (1996) A density-based algorithm for discovering clusters in large spatial databases with noise. Kdd 96:226–231

Giannotti F, Nanni M, Pinelli F, Pedreschi D (2007) Trajectory pattern mining. In: Proceedings of the 13th ACM SIGKDD international conference on knowledge discovery and data mining. ACM, pp 330–339

Kisilevich S, Mansmann F, Keim D (2010) P-DBSCAN: a density based clustering algorithm for exploration and analysis of attractive areas using collections of geo-tagged photos. In: Proceedings of the 1st international conference and exhibition on computing for geospatial research & application. ACM, p 38

Palma AT, Bogorny V, Kuijpers B, Alvares LO (2008) A clustering-based approach for discovering interesting places in trajectories. In: Proceedings of the 2008 ACM symposium on applied computing. ACM, pp 863–868

Rinzivillo S, Pedreschi D, Nanni M, Giannotti F, Andrienko N, Andrienko G (2008) Visually driven analysis of movement data by progressive clustering. Inf Vis 7(3–4):225–239

Chapter 8
Automatic Customization Framework for Efficient Vehicle Routing System Deployment

Jussi Rasku, Tuukka Puranen, Antoine Kalmbach and Tommi Kärkkäinen

Abstract Vehicle routing systems provide several advantages over manual transportation planning and they are attracting growing attention. However, deployment of these systems can be prohibitively costly, especially for small and medium-sized enterprises: the customization, integration, and migration is laborious and requires operations research expetise. We propose an automated configuration workflow for vehicle routing system and data flow customization, which will provide the necessary basis for more experimental work on the subject. Our preliminary results with learning and adaptive algorithms support the assumption of applicability of the proposed configuration framework. The strategies presented here equip implementers with the methods needed, and give an outline for automating the deployment of these systems. This opens up new directions for research in vehicle routing systems, data exchange, model inference, automatic algorithm configuration, algorithm selection, software customization, and domain-specific languages.

8.1 Introduction

Globalization, increased goods consumption, and economic changes pose challenges on transportation logistics. With increasing scale, tightened competition, and environmental concerns, dispatchers stretch their planning capabilities to the limit. Handling all the factors may even be impossible for the human planner (Kleijn 2000), which has spawned an interest in commercial automated route planning systems.

J. Rasku (✉) · T. Kärkkäinen
Department of Mathematical Information Technology, University of Jyväskylä, 40014 Jyväskylä, Finland
e-mail: jussi.rasku@jyu.fi

T. Kärkkäinen
e-mail: tommi.karkkainen@jyu.fi

T. Puranen · A. Kalmbach
NFleet Oy, Pajatie 8 F, 40630 Jyväskylä, Finland
e-mail: tuukka.puranen@nfleet.fi

P. Diez et al. (eds.), *Computational Methods and Models for Transport*, Computational Methods in Applied Sciences 45,
DOI 10.1007/978-3-319-54490-8_8

Combined with the rapid development of IT, this has created a transportation planning tools industry serving operators working with the increasingly complex logistics systems (Hoff et al. 2010).

The advantages of these systems are well known: savings up to 30% in operational costs, reduced planning time, and minimization of human error (Sörensen et al. 2008). Drexl (2011) also note the macroeconomic benefits such as improved traffic flow and lowered emissions. If applied at a large scale, deployment of Vehicle Routing Systems (VRSs) can lead to significant economic and environmental benefits.

A VRS is described in Drexl (2011) as follows: it is an operational planning software that can read in data with vehicles, drivers, depots, customers, and their respective requests connected to geographical locations. This data defines the specific problem scenario. A map view is often used for visual data verification. A VRS then allows manual, interactive or fully automated (optimization-based) construction of routes. The algorithms, that can build a routing plan conforming to the operational rules such as work time regulations, are the key feature of the system. Finally, the system interacts with an existing resource management system, or allows saving and printing the plans for operational use.

The operation environment for a VRS is complex and dynamic (Sörensen et al. 2008; Bräysy and Hasle 2014; Neittaanmäki and Puranen 2015) and poses hard to match requirements for commercial software. In a survey from an industry viewpoint, Hoff et al. (2010) raised a concern that while academic Vehicle Routing Problem (VRP) research has provided efficient algorithms for these problems, they typically use idealized models which omit important facets such as driver breaks, work time regulations, turning restrictions, variation of service times, and congestion. According to Partyka and Hall (2012) the providers are having difficulties in providing holistic solutions due to this complexity.

In addition to shortcomings of the idealized models, different logistic operators have differing requirements (Bräysy and Hasle 2014). As it is not commercially viable to build a unique VRS for each of them (Sörensen et al. 2008) the product is made customizable. Here, a VRS designed for easy deployment needs to capture the features of the common VRP variants. Additionally, solving the problems effectively calls for robustness, adaptivity, and reactivity (Sörensen et al. 2008).

According to Partyka and Hall (2012) routing installations require heavy customization which is mainly done manually. A survey of the Dutch VRS market by Kleijn (2000) agrees: most of the software was at least partially tailor-made. The issue has been identified also in academic research. Puranen (2011) proposes the use of Software Product Lines (SPL) as a mass-customization strategy. It is a promising approach, as these techniques exploit commonalities in a system to effectively manage variation. Applying SPL in other application domains report order of magnitude reductions in time-to-market, engineering overhead, error rates, and cost (Krueger 2002). Preliminary results in (Puranen 2011) suggest that these benefits are achievable also with VRSs. The extended rationale for the work, as presented in (Neittaanmäki and Puranen 2015), is that the underutilization of route optimization is not due to the shortcomings of models and algorithms, but due to problems in deployment.

The challenges in implementation and deployment call for an approach that could forward the adoption of route optimization. In this paper, we propose such an approach as a set of actions and techniques to automate the flow of data through a VRS. Acting upon presented ideas allows utilization of the recent advances in Software Engineering, Machine Learning, and Operations Research.

In related works Desrochers et al. (1999) describe a VRS that could be used by a consultant with a basic understanding of mathematical programming. Similarly, Maturana et al. (2004) describe a decision support system generator that substantially lowers the cost of developing such systems, although in their solution the model and data structures has to be still defined manually. Also, Hoff et al. (2010) envision a tool exhibiting some of the properties presented here. Despite these ideas, a planning and decision support system that allows as extensive automation as ours has not been previously described. No customization framework for the needed automation methods has existed, and its interaction within VRSs has not been previously explored. In this paper, we address this by providing an automated configuration workflow for VRSs and a review on the automation methods for different phases of the process. The customization framework should be of interest not only to operations researchers, but also to providers of VRSs.

For an overview, Sect. 8.2 provides a review of the trends in vehicle routing research. In Sect. 8.3 we recognize the opportunities for automation in customizing from data flow perspective. Section 8.4 we present our proposition for solving some pressing problems in VRS deployment and Sect. 8.5 reviews our preliminary experimental results. In Sect. 8.6 we conclude our study.

8.2 Trends in Vehicle Routing Problem Research

VRP has been under intensive research ever since was first introduced by Dantzig and Ramser (1959). VRP concerns the task of finding optimal *routes* for a *fleet* of *vehicles* leaving typically from a *depot* to serve a specified number of *requests*. Objectives can be anything from minimizing the number of vehicles or total travel distance to complex multiobjective business goals. Over 50 years of academic interest has experienced many shifts of research focus. The trends in VRP research, as recognized e.g. by Puranen (2011), are illustrated in Fig. 8.1.

The early models were **idealized**, partly due to the limitations of computational hardware and solution methods of the time. Since the early days, the trend has been towards more complex and more realistic **"rich"** problems (Sörensen et al. 2008; Hoff et al. 2010). Rich models extend the classical formulation with complex decisions and objectives as well as can introduce many operational constraints. In addition, a

models:	idealized	→	rich	→	unified	→	inferred
methods:	simple	→	refined	→	adaptive	→	learning

Fig. 8.1 Trends in vehicle routing research. Adapted from (Puranen 2011)

number of new aspects have been proposed; for example Hoff et al. (2010) call for more explicit handling of stochasticity and risk in the models, and stress the need for research on real time and dynamic routing. For reviews on rich VRP research in the context of commercial solutions, we refer to Hasle and Kloster (2007), Drexl (2011), Bräysy and Hasle (2014). Recently, there has been efforts to develop **unified** modeling approaches with generic and flexible modeling structures that can capture the aspects of different VRP variants (Ropke and Pisinger 2006; Hasle and Kloster 2007; Irnich 2008; Welch et al. 2011; Puranen 2012). Vidal et al. (2012) make a synthesis on previous research and propose a naming scheme for these variants. Unified modeling frameworks often provide a Domain Specific Language (DSL) for describing the problems. One of the contributions of this paper concerns the rightmost transition in modeling: we argue that the advances in unified modeling enable model **inference** where *composite* optimization model can be automatically or semi-automatically formed by inferring the composition of the model from the problem instance data.

The solution methodologies have followed a similar trend. The first methods relied on **simple** heuristics or on mathematical programming as in the original paper by Dantzig and Ramser (1959). The growing problem size and model complexity led to interest in more **refined** and sophisticated methods. However, due to scale of the real-life problems, exact solution methods cannot always be used. Thus, a number of heuristics and metaheuristics have been proposed. For surveys in solving VRPs, see, e.g., Toth and Vigo (2002), Cordeau et al. (2005), Laporte (2007). Recently, there has been interest in **adaptive** and self-adjusting methods where algorithms observe the optimization progress and react accordingly. This trend was recognized, for example, in the survey by Vidal et al. (2013). A newer trend is the application of **learning** hyperheuristic systems, which involve using data-driven techniques that enable and disable different algorithms depending on the observed search space. This involves identifying situations similar to those found in the history data or knowledge model. For a survey on using hyperheuristics in combinatorial optimization we refer to Kotthoff (2014).

The disadvantage of unified and "rich" models and refined versatile solving methods is that they may make the deployment more complicated (Neittaanmäki and Puranen 2015). Also, note that in most of the case studies in the aforementioned surveys, the derivation of the model, selection of the algorithms, and fine tuning of the methods is done manually by the researchers based on their expertise. Unfortunately, this does not scale in a commercial setting and poses a barrier for the deployment of VRSs.

8.3 Data Flow in a Vehicle Routing System

In this section, we outline the *data flow* through a VRS, or more specifically, how the problem instance is passed from system module to another. The flow of information is one of the main aspects affecting the deployment, integration, and utilization of

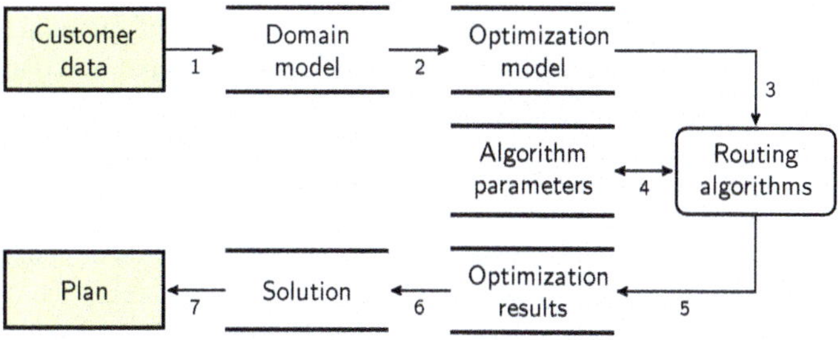

Fig. 8.2 Data flow of a problem instance trough a VRS

the system. Describing the modular structure of a typical VRS in detail is omitted, and the reader is referred to Drexl (2011), Puranen (2011), Bräysy and Hasle (2014).

The data flow can be divided into phases as illustrated in Fig. 8.2. First, the data is read from a data storage, such as files or relational database, and then constructed into a domain model (1). Domain model offers primitives for concepts such as truck, driver, and request. The domain model is then translated into optimization model (2). This involves describing the decision variables, the objective function, and the necessary constraints. Note that a DSL or similar technique has to be used to capture the aspects of the specific routing problem. Result of this transformation is a mathematical optimization model that can be then completed with the problem instance specific variable values. The modeled problem can now be fed to the routing algorithms residing in the solver module (3). Effectiveness of the algorithms depends heavily on the algorithm parameters (Hoos 2012). Thus, when adapting a VRS care is needed to derive a set of suitable parameter values (4). After the optimization (5) the results can be transformed back to the primitives of the domain model (6) which in turn are translated into an actionable plan (7).

Provided that the VRS is generic enough to model a wide set of different and "rich" VRP variants, and that it includes a set of modern metaheuristics and local search based routing algorithms, the biggest effort in adopting a VRS for a new customer is to make sure that the data is read and processed correctly (Neittaanmäki and Puranen 2015). VRS providers have several options to manage the data flow:

(a) Force an identical data flow for all customers. This will remove much of the flexibility and only a narrow set of problem types can be efficiently addressed by the VRS.
(b) Customize the data flow manually on a case-by-case basis. Here multiple model variants and use cases can be supported, but the customization heavily depends on manual work and expertise.
(c) Outsource the customization to a third party or to the customer by providing a way to externally configure the system via, e.g. a DSL. The challenge is to provide enough training and sufficient tools for the third party.

(d) Automate the customization so that fixing any given set of functionality inside the VRS is done automatically based on the customer input and data. In addition, if the provider has access to the history data during the customization, the automation might even be learning, that is, with every new modeled problem and deployment the software gains experience.

Designing the software in a way that the flexibility is maximized makes the system applicable in larger number of different contexts, thus making the approaches (c) and (d) feasible. The challenge for (c) is that many logistic operators are small, and lack the necessary expertise to understand the inherent complexity in selecting, configuring and deploying VRSs (Neittaanmäki and Puranen 2015). Therefore, out of these, the automation based approach (d) is the one that is more scalable and cost-effective. This validates the need for the proposed customization framework.

Each of the data conversion phases Fig. 8.2 expose a potential point of customization. In practice, this *variability* is exposed by configurable behavior of the software system, and it needs to be managed somehow. From a theoretical perspective, this has been addressed by the techniques in the area of software product line engineering (SPLE) (Pohl et al. 2005). In SPLE, the developed system is divided into two layers: domain layer and application layer. The domain layer of the system captures the generic properties of the current domain, and the application layer is used to define customized application instances with *variability points*. It is a predefined point in the system, in which variation between the applications occur (Jacobson et al. 1997). The specialized expertise required in the customization of VRSs prohibits manually managed mass customization. Instead, we suggest the use of machine-learning based adaptive mass customization techniques, and argue that these represent one of the key technologies in achieving cost-effective routing system deployment.

8.4 The Automation and Customization Framework

Our main contribution is an outline, or a vision, of how highly automated and easy-to-deploy VRS could be constructed. This customization framework could also enable experimentation with different automation approaches, but here we concentrate on the techniques we have either successfully applied ourselves, or see as pragmatic solutions to the presented opportunities for workflow automation. We limit ourselves to well-known methods used in related fields, and assume a generic solver module capable of expressing a wide set of "rich" VRP variants. The section follows the structure of Fig. 8.2 with each phase having a corresponding subsection.

8.4.1 Interpreting Customer Data

Input data → Domain model

Interpreting the customer data and transforming it into routing problem starts with the creation of a domain model, which represents the real world entities that form the routing problem. The transformation task consists of taking the problem data as an input and then extracting the data into the domain model. In the simplest case, one can specify a data format that is required and the VRS simply parses this data into a model. This becomes problematic one needs to support different formats, as maintaining several many-to-one mappings can quickly become an onerous task.

A likely scenario for data integration is a relational dataset, such as relational database, but in general, any kind of flat dataset with interconnected files can be used. To illustrate, one part of the dataset could consist of ordinary files that pertain to drivers and vehicles, and the other deliveries and locations. Finding semantic links between the relations in these datasets is what we call *join inference*, which in turn is based on foreign key discovery (Acar and Motro 2009; Rostin et al. 2009). We propose join inference as a model that can learn the semantic links between a set of relations. It is used to produce a cohesive union of data, the joined relation.

After join inference has been done, we propose the use of schema mapping (Bellahsene 2011) to extract the required information from the data. Schema mapping consists of finding pairings between two schemas. A *schema* is a formal description of the information contained in a relation; crudely, this would be a set of data attributes, or column headers. Having to find these attribute pairings makes the problem a kind of *data exchange problem* (Kolaitis 2005), where the goal is to take data from different sources and assimilate it, in this context, to the domain model of a VRS.

8.4.2 Inferring the Optimization Model

Domain model → Optimization model

After mapping the input data to the domain model, it must be translated into a format understood by the VRP solver. This includes choosing an optimization model. We were unable to find related work on automating this step. Therefore, we proceed to propose four approaches for implementing such an automated transformation:

1. **Separate models:** methods from Sect. 8.4.1 can be used on domain model to map it against a selection of optimization models. Out of these, the one with the best fit is selected and completed with instance data. This is suitable approach only if a VRS has support for a limited number of VRP variants.

2. **Coupled models:** a number of domain and optimization models are coupled together with predefined pairwise transformations. Data interpretation from Sect. 8.4.1 is done with all domain models in the coupling set and then the one with the best schema mapping (along some criteria) is selected. This has the same constraints as the previous approach.
3. **Model composition:** the optimization model is composed of different objects that may correspond to partial objectives, decisions, or constraints. Filled domain model is matched against each optimization model component and if a threshold is crossed, the component is included to the composite model.
4. **Model reduction:** alternatively, the initial optimization model may be "complete" or unified in a sense that is capable of expressing all the supported VRP features. After doing schema mapping between the domain and optimization models, the unused elements, for which the variable values were not set, are removed from the optimization model.

Besides domain model, other sources for deducing the optimization model include e.g. the vocabulary used in the data. To illustrate, a data field revealing that the transportation involves people, refers to use of a dial-a-ride optimization model. The unified naming convention for VRP variants in Vidal et al. (2013) might prove to be useful in recognizing the different modeling constructs for the model inference. We note that the feasibility of applying automation in this phase is uncertain, mostly because of the lack of prior published research on the topic.

8.4.3 Selecting the Suitable Optimization Algorithms

Optimization model → Algorithm performance predictions

Industrial solutions tend to favor algorithms based on heuristics (Sörensen et al. 2008; Bräysy and Hasle 2014), and many implement a collection of different algorithms to gain extra flexibility. It is also known, that the performance of an algorithm varies greatly between different routing variants and even problem instances (Hoos 2012). Therefore, it is important to use an algorithm that is efficient in solving the given problem. Portfolio-based algorithm selection techniques such as SATzilla (Xu and Leyton-brown 2008) use statistical models to select the algorithm for solving a given problem instance. In VRS this approach could be applied to select the higher level algorithmic components: a metaheuristic could be selected based on the instance characteristics and predicted performance.

Another way to improve solver performance is the utilization of so called *hyper-heuristics*. Instead of using a single algorithm or a manually constructed combination, a hyperheuristic acts as a high level learning "supervisor" algorithm that *selects and combines* lower level algorithms from a portfolio on the fly.

Similar ideas have been tried in VRP, for example, by using several simple heuristics in varied order to escape local optima. Pisinger and Ropke (2007) proposed a method, where adaptive heuristic selection is done among intensification and diversification heuristic operators. Garrido has proposed the use of hyperheuristics to select local search operators in solving different VRP variants (Garrido and Riff 2010). VRP was also one of the problem domains in Cross-domain Heuristic Search Challenge (CHeSC2011) where a number of domain independent hyperheuristics were evaluated (Walker et al. 2012).

We note that these schemes should be useful when adapting an industrial VRS to a new set of end user provided sample problem instances. Our experimental work to explore these possibilities is in preparation.

8.4.4 Configuring the Optimization Algorithm

Optimization model and Observed performance → Algorithm parameter values

The algorithms used to solve hard computational problems often reveal parameters that can be used to change the behaviour of the algorithm and adapt it into solving a specific problem instance (Hoos 2012). The settings of the algorithm parameters have a substantial effect on the performance of the algorithm. However, setting them manually is a non-trivial task requiring expertise and effort trough experimentation (Hoos 2012). Therefore, automatic search approaches have been proposed to what is in literature known as the problem of *automated algorithm configuration (AAC)*.

In practice, AAC can be used to automatically adapt the a routing solver for each VRS deployment. This allows the VRS provider to get the best performance out of the implemented algorithms. Also, after enough experimentation, archetypes of routing problems might emerge. With this history data the previous configuration effort could be reused to provide more varied algorithm defaults for the solver. In fact, several AAC methods have proven successful with VRP (see e.g. Pellegrini and Birattari 2006). Of particular interest in this context is the work in Becker et al. (2006), where they tuned the parameters of a commercial VRP solver with real-world routing problem instances. Our recent experimental work (Rasku et al. 2014) verifies this and gives suggestions on which AAC methods to use to configure VRP metaheuristics.

Current state-of-the-art methods like SMAC from Hutter et al. (2011) or I/F-Race from Balaprakash et al. (2007) support all parameter types, are able to use extra information like the parameter structure, interactions or hierarchies, and use several intensification techniques that aim to save computationally expensive parameter configuration evaluations. The benefits of can be striking: Hutter et al. (2010) were able to achieve up to 50-fold speedup over the default parameters of the CPLEX solver.

The main challenge of applying AAC in routing, however, is that the runtime on real world routing cases may be hours, especially in presence of complex constraints (Becker et al. 2006). Fortunately, the focus of the AAC research has been recently shifting onto overcoming these challenges, see e.g. Mascia et al. (2013).

8.4.5 Solving the VRP Problems

Optimization model and Algorithms and their parameter values → Optimization results

The solver module is responsible of performing the optimization, which contains the tasks of mapping of tasks to vehicles as well as routing the vehicles as efficiently as possible according to the objective function. The search is performed until a predefined stopping criterion has been met, or the user ends the process.

From the process perspective, the ability to predict and adjust the runtime is a major concern. The same system may be operated under tight time-constraints for planning, whereas some users prefer the added robustness of a more thorough search. It is probable that this variability is exposed e.g. as stopping criteria.

Another viewpoint to solver module customization is the availability of computational resources. In many cases, the routing system is still run in a desktop environment, but increasingly, optimization services are available through cloud services (Bräysy and Hasle 2014). This opens a new dimension in the customization, namely the allocated computing time, resources and priority based on the customer requirements, service level agreements, and instance characteristics, which all adds in to the complexity of deploying the system.

8.4.6 Interpreting the Optimization Results

Optimization model and Optimization results → Domain model (solution)

The optimization solver module usually returns the resulting plan in the mathematical format it uses internally. The interpretation of the optimization results is linked to the construction of the optimization model. Whereas in model construction the decision variables are selected based the data in the domain model, in this phase the values of the decision variables need to be interpreted back to the relations and values of the entities in the domain model. We can use an inverse transformation of the one in Sect. 8.4.2 to decode the solution.

One issue in the interpretation of the results is the type of the decoding. It may be that the decoding is not one-to-one. That is, there may be multiple possible plans the optimized solution can be decoded to. For example, in a classical VRP all the trucks are identical and it does not matter how the vehicles and routes are mapped Puranen (2011). This potential unambiguity may have undesired consequences if it is not taken into account.

8.4.7 Producing a Formatted Plan

Domain model (solution) → Output plan

Ultimately the user of a VRS needs to apply the plan into practice. Different users have different formats, output data requirements, and reporting needs, so in the final data transformation step an automated VRS could adapt its output to the format most convenient to the end user.

If the interfaced system includes plan generation, it could be enough to do the schema mapping procedure from Sect. 8.4.1 in reverse. The existing system would then compose the output document to that is to be handed to the drivers. Another option is to infer the structure of an example document using methods such as table extraction, visual object and information extraction, and entity identification (Lin et al. 2006; Krüpl-Sypien et al. 2011). This produces a template which then can be filled with the relevant data from the solver. Similar technique has been used, for example, in web page content and structure extraction to reformat the web page content for mobile clients (Krüpl-Sypien et al. 2011).

8.5 Preliminary Experimental Results

To demonstrate the potential of automatic configuration of route optimization algorithms, we configured the three metaheuristics (Record-to-Record travel *RTR*, simulated annealing *SA*, and ejection *EJ*) provided by the VRPH library (Groër et al. 2010) on four real world based benchmark instances from the literature. For details of the experimental setup see Rasku et al. (2014).

The target problem instances were `F-n45-k4`, `F-n72-k4`, and `F-n135-k7` from (Fisher 1994) with 45, 72 and 135 requests and the 385 request instance `tai385` from Taillard (1993). The `tai385` instance is generated based on the locations and census of population data from canton of Vaud in Switzerland, whereas the instances `F-n45-k4` and `F-n135-k7` are from a day of grocery deliveries from the Ontario terminal of National Groceries Limited. The `F-n72-k4` instance data is obtained from Exxon associated case involving the delivery of tires, batteries and other accessories to gas stations. We used SMAC (Hutter et al. 2011) (version 2.3.5) and Iterated F-Race (Balaprakash et al. 2007) implementation described in (López-Ibánez et al. 2011) (version 1.0.7) and restricted to evaluation budget of 500 invocations. Each metaheuristics was configured separately for each of the target problem instance. A 30 second cutoff was used for the solvers.

Results of the configuration runs are presented in Table 8.1. On average, the quality of the results was improved by 1.65% points with the use of AAC, which means that the relative optimality gap was closed by 73%. Furthermore, the performance of the metaheuristics was more consistent when configured, as can be observed from the standard deviations.

Table 8.1 Average AAC results for all solver-instance pairs. Results are given as percentage from the best known solution (relative optimality gap). Statistically better ($p < 0.05$) results are bolded, evaluation budget violations of more than 5% are in italics, and the standard deviation over 100 VRP solutions is given in parentheses

Target	Quality on the target instance			Quality on the other instances		
	Defaults	I/F-Race	SMAC	Defaults	I/F-Race	SMAC
F-n45 EJ	0.49 (0.35)	**0.12 (0.23)**	**0.15 (0.25)**	2.57 (2.19)	**2.21 (1.41)**	2.70 (2.07)
F-n45 RTR	0.48 (0.40)	0.01 (0.04)	**0.00 (0.00)**	11.25 (0.40)	**5.32 (3.01)**	**6.02 (3.56)**
F-n45 SA	0.30 (0.34)	**0.03 (0.14)**	**0.01 (0.09)**	8.91 (1.54)	**6.55 (4.97)**	**7.68 (6.70)**
F-n72 EJ	0.99 (2.15)	**0.19 (1.03)**	**0.16 (1.11)**	**1.98 (0.54)**	2.15 (0.88)	2.11 (0.82)
F-n72 RTR	6.63 (0.28)	**0.00 (0.00)**	**0.00 (0.00)**	4.94 (0.51)	**3.86 (1.01)**	**3.66 (1.02)**
F-n72 SA	3.80 (1.75)	**0.05 (0.15)**	**0.02 (0.09)**	5.06 (0.52)	**2.66 (1.11)**	**3.06 (1.41)**
F-n135 EJ	0.24 (0.29)	**0.19 (0.28)**	**0.17 (0.15)**	2.96 (2.58)	2.01 (1.61)	**1.88 (0.92)**
F-n135 RTR	1.62 (0.07)	0.06 (0.08)	**0.02 (0.03)**	9.94 (0.57)	**4.71 (3.00)**	5.65 (2.17)
F-n135 SA	0.11 (0.07)	*0.14 (0.14)*	**0.08 (0.06)**	8.99 (1.57)	***6.42 (3.65)***	**6.25 (2.89)**
tai385 EJ	1.23 (0.28)	1.10 (0.23)	**1.02 (0.18)**	1.92 (2.46)	**0.72 (0.53)**	**0.76 (0.43)**
tai385 RTR	2.91 (0.27)	1.00 (0.22)	**0.88 (0.18)**	8.61 (0.48)	3.99 (2.09)	**3.47 (1.85)**
tai385 SA	4.67 (0.40)	***1.04 (0.24)***	1.18 (0.27)	3.74 (2.06)	***2.15 (2.51)***	5.79 (4.63)

Because the metaheuristics were configured for each instance separately, we acknowledge the danger of overtuning (Hutter et al. 2010). To observe the effect, the rightmost three columns of Table 8.1 present the performance of the resulting parameter configurations on the other three remaining instances. These columns can be interpreted as the result of a 4-fold cross-validation. The configurators overfit only for the `F-n72-k4`, and for all other targets the solution quality is statistically significantly improved on average by 2.4% points (a 37% improvement). The solver behavior becomes slightly more erratic as can be perceived from the standard deviations. However, this is likely to be a byproduct of the improved solution quality and the more rugged fitness landscape of a multi-instance problem set. As suggested by the results in (Rasku et al. 2014), if tuned on the entire instance set, the robustness of the solvers on similar instances is expected to improve.

Our proposed solution to increase the level of automation in the data import phase is presented in (Kalmbach 2014). To summarize, Kalmbach (2014) provided a formulation for the data import and model inference problem, presented a decision trees (Quinlan 1986) based approach for join inference and schema mapping, and explored its applicability in importing of schemaless routing instance data. Two decision trees for *capacity* and request *location* mapping are provided in Fig. 8.3 as an illustration of the generated inference rules. The proof-of-concept tool is able to recognize the nature of each column in a column-oriented input for the generated test data, and is thus capable of generating simple mapping rules between input and the domain model.

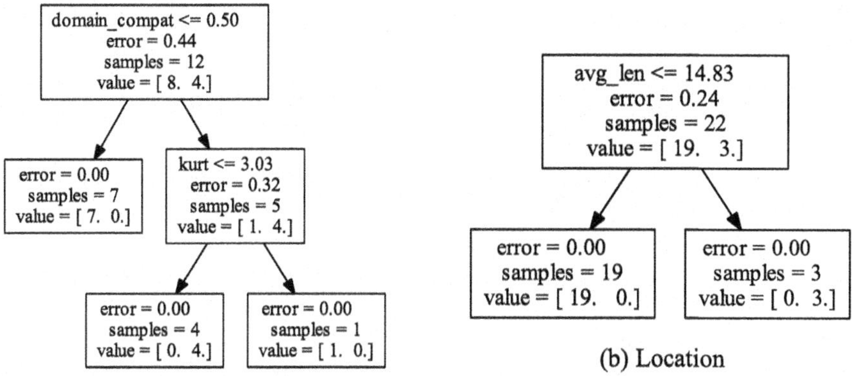

(a) Capacity

(b) Location

Fig. 8.3 Decision trees for the schema mapping of two domain model attributes

8.6 Conclusions

Vehicle routing systems provide several advantages over manual transportation planning, but the deployment of these systems is in many cases laborious and costly. In addition, migration from the current system with associated customization and integration challenges create practical obstacles that prevent the latest advances in operations research from being disseminated to wide use. The focus in academic research is in modeling and solving efficiency whereas in commercial routing systems usability, flexibility, and scalability are more important. Tighter interaction between the two is needed in order to effectively solve real-world routing problems (Bräysy and Hasle 2014).

The advances in technologies such as GPS and RFID, and drop in data warehousing prices, have made transportation big data collection possible and economical. Concurrently, logistics operators have begun to see the information as a vital asset that can be used in decision making. This opens new possibilities for machine learning, for which the accuracy is dependent of the amount and availability of data that can be used to train the models. Therefore, these trends have paved the road for a new generation of vehicle routing systems that can utilize machine learning to automate the customization and deployment. This in turn has the potential to increase the effectiveness and robustness as the system can be adapted automatically to the particularities of a problem instance. The goal is to diminish the importance of an operations researcher in the deployment process and consequently to permit higher scalability and more widespread deployment of route optimization.

In this paper, we have outlined a customization framework for the automation of data transformation operations inside a routing system. Our framework recognizes seven transformation steps, each open for system customization. We also provide suggestions on automating these steps. Our preliminary empirical results are promising, but further experimental work is required to establish whether all the proposed techniques are fully applicable in practice.

To evaluate the proposed customization framework, we reflect it against the framework for analyzing VRS deployment published in (Neittaanmäki and Puranen 2015). Neittaanmäki and Puranen (2015) recognize several practical adoption and deployment barriers for the VRSs. They see the involvement of an optimization expert as a prohibiting investment and call for an increased automation of the deployment process. Their deployment process is split into three phases: data, process, and system integration. To see in which extent our proposed customization framework can resolve the 18 barriers they recognized, we proceed to give some possible solutions to the recognized issues: In data integration step, the missing, low quality and incomplete data could be automatically imputed, or at least recognized with machine learning. The data structure inference from Sect. 8.4.1 can help when acquiring and combining the data from existing systems. In addition, because of the techniques proposed in Sect. 8.4.2, it takes less expertise to generate the optimization model. As demonstrated by our experiments, the plan quality can be improved, sometimes significantly, using automatic algorithm configuration (Sect. 8.4.4). Use of automation results into lower perceived complexity and improved usability that can instill trust in the users to the system and to the plans it generates. On the system integration level, the automation makes integration easier and faster, which in turn can make the system deployment cheaper, less dependent on expertise and other resources, and flexible to the current and future changes in operations.

Taken together, we argue that in order to bring the latest academic routing knowledge to the hands of logistics operators in a massive scale, the automatic configuration approach, as presented in this paper, is needed. The recent trends in VRP research seem to converge towards generic reusable modeling and highly adaptive and configurable modeling frameworks, but we have shown that several other areas in practical system integration and deployment need to be considered in order to effectively apply these into practice. This requires extensive further studies in several disciplines, but should provide a promising area of research with a potential for a wide array of practical benefits.

References

Acar AC, Motro A (2009) Efficient discovery of join plans in schemaless data. In: Proceedings of the 2009 international database engineering & applications symposium, ACM, New York, NY, USA, IDEAS 2009, p 111

Balaprakash P, Birattari M, Stützle T (2007) Improvement strategies for the F-Race algorithm: sampling design and iterative refinement. Technical report, IRIDIA, Université Libre de Bruxelles

Becker S, Gottlieb J, Stützle T (2006) Applications of racing algorithms: an industrial perspective. In: Proceedings of the 7th international conference on artificial evolution, EA 2005. Springer, Heidelberg, pp 271–283

Bellahsene Z (2011) Schema matching and mapping. Springer, Heidelberg

Bräysy O, Hasle G (2014) Software tools and emerging technologies for vehicle routing and intermodal transportation, SIAM, Chap 12, pp 351–380. MOS-SIAM Series on Optimization
Cordeau JF, Gendreau M, Hertz A, Laporte G, Sormany JS (2005) New heuristics for the vehicle routing problem. In: Logistics systems: design and optimization, Chap 9. Springer, New York, pp 279–297
Dantzig GB, Ramser JH (1959) The truck dispatching problem. Manag Sci 6(1):80–91
Desrochers M, Jones CV, Lenstra JK, Savelsbergh MWP, Stougie L (1999) Towards a model and algorithm management system for vehicle routing and scheduling problems. Decis Support Syst 25(2):109–133
Drexl M (2011) Rich vehicle routing in theory and practice. Technical report 1104, Gutenberg School of Management and Economics, Johannes Gutenberg University Mainz
Fisher ML (1994) Optimal solution of vehicle routing problems using minimum k-trees. Oper Res 42(4):626–642
Garrido P, Riff MC (2010) DVRP: a hard dynamic combinatorial optimisation problem tackled by an evolutionary hyper-heuristic. J Heuristics 16(6):795–834
Groër C, Golden B, Wasil E (2010) A library of local search heuristics for the vehicle routing problem. Math Program Comput 2(2):79–101
Hasle G, Kloster O (2007) Industrial vehicle routing. In: Geometric modelling, numerical simulation, and optimization. Springer, Heidelberg, pp 397–435
Hoff A, Andersson H, Christiansen M, Hasle G, Løkketangen A (2010) Industrial aspects and literature survey: fleet composition and routing. Comput Oper Res 37(12):2041–2061
Hoos HH (2012) Automated algorithm configuration and parameter tuning. In: Autonomous search. Springer, Heidelberg, pp 37–71
Hutter F, Hoos HH, Leyton-Brown K (2010) Automated configuration of mixed integer programming solvers. CPAIOR, Lecture Notes in Computer Science, vol 6140. Springer, Heidelberg, pp 186–202
Hutter F, Hoos H, Leyton-Brown K (2011) Sequential model-based optimization for general algorithm configuration. In: Learning and intelligent optimization. Springer, Heidelberg, pp 507–523
Irnich S (2008) A unified modeling and solution framework for vehicle routing and local search-based metaheuristics. Informs J Comput 20(2):270–287
Jacobson I, Griss M, Jonsson P (1997) Software reuse: architecture, process and organization for business success. ACM Press/Addison-Wesley Publishing Co., New York
Kalmbach A (2014) Fleet inference : importing vehicle routing problems using machine learning. Master's thesis, University of Jyväskylä, Department of mathematical information technology
Kleijn MJ (2000) Tourenplanungssoftware: ein vergleich für den niederländischen markt. Internationales Verkehrswesen 52(10):454–455
Kolaitis PG (2005) Schema mappings, data exchange, and metadata management. In: Proceedings of the twenty-fourth ACM SIGMOD-SIGACT-SIGART symposium on principles of database systems. ACM, New York, NY, USA, PODS 2005, p 6175
Kotthoff L (2014) Algorithm selection for combinatorial search problems: a survey. AI Mag 35(3):48–60
Krueger CW (2002) Easing the transition to software mass customization. In: Linden F (ed) Software product-family engineering, Lecture Notes in Computer Science, vol 2290. Springer, Heidelberg, pp 282–293
Krüpl-Sypien B, Fayzrakhmanov RR, Holzinger W, Panzenböck M, Baumgartner R (2011) A versatile model for web page representation, information extraction and content re-packaging. Proceedings of the 11th ACM symposium on document engineering. ACM, New York, pp 129–138
Laporte G (2007) What you should know about the vehicle routing problem. Naval Res Logistics 54(8):811–819
Lin X, Hui C, Nelson G, Durante E (2006) Active document versioning: from layout understanding to adjustment. In: Taghva K, Lin X (eds) Proceedings of document recognition and retrieval XIII, SPIE, vol 6067

López-Ibánez M, Dubois-Lacoste J, Stützle T, Birattari M (2011) The irace package, iterated race for automatic algorithm configuration. Technical report, IRIDIA, Université Libre de Bruxelles
Mascia F, Birattari M, Stützle T (2013) An experimental protocol for tuning algorithms on large instances. In: Learning and intelligent optimization. Springer, Heidelberg
Maturana S, Ferrer JC, Barañao F (2004) Design and implementation of an optimization-based decision support system generator. Eur J Oper Res 154(1):170–183
Neittaanmäki P, Puranen T (2015) Scalable deployment of efficient transportation optimization for smes and public sector. Advances in evolutionary and deterministic methods for design, optimization and control in engineering and sciences. Springer, Heidelberg, pp 473–484
Partyka J, Hall R (2012) Software survey: vehicle routing. OR/MS Today, 39(1)
Pellegrini P, Birattari M (2006) The relevance of tuning the parameters of metaheuristics. Technical report, RIDIA, Université Libre de Bruxelles
Pisinger D, Ropke S (2007) A general heuristic for vehicle routing problems. Comput Oper Res 34(8):2403–2435
Pohl K, Böckle G, van der Linden FJ (2005) Software product line engineering: foundations principles and techniques. Springer, Heidelberg
Puranen T (2011) Metaheuristics meet metamodels—a modeling language and a product line architecture for route optimization systems. Ph.D thesis, University of Jyväskylä, Jyväskylä studies in computing, 1456–5390, 134
Puranen T (2012) Producing routing systems flexibly using a VRP metamodel and a software product line. In: Proceedings of operations research 2011. Springer, Heidelberg, pp 407–412
Quinlan JR (1986) Induction of decision trees. Mach Learn 1(1):81–106
Rasku J, Musliu N, Kärkkäinen T (2014) Automating the parameter selection in VRP: an off-line parameter tuning tool comparison. In: Modeling, simulation and optimization for science and technology, computational methods in applied sciences, vol 34. Springer, Heidelberg, pp 191–209
Ropke S, Pisinger D (2006) A unified heuristic for a large class of vehicle routing problems with Backhauls. Eur J Oper Res 171(3):750–775
Rostin A, Albrecht O, Bauckmann J, Naumann F, Leser U (2009) A machine learning approach to foreign key discovery. In: 12th international workshop on the web and databases (WebDB)
Sörensen K, Sevaux M, Schittekat P (2008) Multiple neighbourhood search in commercial VRP packages: evolving towards self-adaptive methods. In: Adaptive and multilevel metaheuristics, Springer, Heidelberg, pp 239–253
Taillard E (1993) Parallel iterative search methods for vehicle routing problems. Networks 23(8):661–673
Toth P, Vigo D (eds) (2002) The vehicle routing problem. SIAM
Vidal T, Crainic TG, Gendreau M, Prins C (2012) A unified solution framework for multi-attribute vehicle routing problems. Technical report, CIRRELT
Vidal T, Crainic TG, Gendreau M, Prins C (2013) Heuristics for multi-attribute vehicle routing problems: a survey and synthesis. Eur J Oper Res 231(1):1–21
Walker JD, Ochoa G, Gendreau M, Burke EK (2012) Vehicle routing and adaptive iterated local search within the hyflex hyper-heuristic framework. In: 6th international conference on learning and intelligent optimization, Lecture Notes in Computer Science, vol 7219. Springer, Heidelberg, pp 265–276
Welch PG, Ekárt A, Buckingham C (2011) A proposed meta-model for combinatorial optimisation problems within transport logistics. In: MIC 2011: The IX metaheuristics international conference, vol IX
Xu L, Leyton-brown K (2008) SATzilla: portfolio-based algorithm selection for SAT. Artif Intell 32:565–606

Chapter 9
The Multi-period Fleet Size and Mix Vehicle Routing Problem with Stochastic Demands

Urooj Pasha, Arild Hoff and Lars Magnus Hvattum

Abstract The multi-period fleet size and mix vehicle routing problem with stochastic demands is a new optimization problem that arises from the need to make strategic fleet sizing decisions while taking into consideration tactical planning and operational uncertainty. The setting is a distribution company that delivers goods to a set of customers and where the expected demand for different customers vary from period to period. The actual demand in a given period is stochastic, and is only revealed when visiting the customer. The objective is to minimize total expected costs, consisting of vehicle acquisition costs, routing costs, and the expected cost of route failures. Route failures occur when a vehicle arrives at a customer with an insufficient amount of goods, resulting in the need to refill the vehicle at the depot. The problem is formulated as a mixed integer programming problem. A heuristic for solving the problem is described and implemented, and a computational study is conducted on a set of varied test instances.

9.1 Introduction

The efficient transportation of goods is essential for most companies, and transportation costs can account for 20% of the total cost of a product (Hoff et al. 2010). Making proper decisions regarding transportation requires information about available vehicle types, their cost and capacity, the set of customers to visit and their corresponding demand. In a deterministic case, all information is available at the time of planning. However, in real life, this is rarely the case and there will be uncertainty regarding information that is only revealed after the main decisions have been made.

U. Pasha · A. Hoff (✉) · L.M. Hvattum
Molde University College, P.O. Box 2110, 6402 Molde, Norway
e-mail: arild.hoff@himolde.no

U. Pasha
e-mail: uroojpasha@hotmail.com

L.M. Hvattum
e-mail: lars.m.hvattum@himolde.no

P. Diez et al. (eds.), *Computational Methods and Models for Transport*,
Computational Methods in Applied Sciences 45,
DOI 10.1007/978-3-319-54490-8_9

This chapter considers situations where a transporter needs to determine a suitable fleet size and mix for servicing regular customers. Customers have demands that vary between periods, for example from one day of the week to another, and vehicles are not obliged to follow the same route in every period. For a given period, the actual demand of each customer is stochastic, and is revealed only when the customer is first visited. If a vehicle arrives at a customer with an insufficient amount of goods, it must return to the depot to refill before continuing the service.

Yang et al. mention several practical distribution settings where stochasticity and refilling can occur, such as the transportation of beer to retail outlets, the distribution of baked goods to food stores, and stocking of vending machines. In these situations, distributors fill up their trucks and visit customers with the intent of replacing depleted stock, and the expected demand vary for different days of a week (Yang et al. 2000). Additional examples, with materials being collected rather than delivered, include trash collection and sludge disposal, where the amount of goods is uncertain but can have a higher or lower expected value on certain days.

To handle these situations, the *multi-period fleet size and mix vehicle routing problem with stochastic demands* (MPFSMVRPSD) is formally defined in this chapter. The MPFSMVRPSD is related to the *fleet size and mix vehicle routing problem* (Golden et al. 1984), the *period vehicle routing problem* (Francis et al. 2008), and the *vehicle routing problem with stochastic demands* (Dror et al. 1989).

The remainder of this chapter is structured as follows. A brief overview of related literature is given in Sect. 9.2. Section 9.3 contains a detailed problem description and a mathematical model. Heuristics for solving the problem are described in Sect. 9.4. A computational study is described in Sect. 9.5, and concluding remarks are given in Sect. 9.6.

9.2 Literature Review

In the following, we discuss variants of the *vehicle routing problem* (VRP) that are related to the MPFSMVRPSD. In the *fleet size and mix VRP* (FSMVRP) described by (Golden et al. 1984), an unlimited number of vehicles with different capacity and cost are available. The objective is to find the optimal composition of a fleet of vehicles, combined with the best possible routing to visit a set of customers, where both fixed vehicle acquisition costs and variable routing costs are considered. Each route should start and end at the depot, and the total demand of the customers in a route should not exceed the vehicle capacity. The problem considers a specific time period where the demands of the customers are fixed and known.

As an extension of the FSMVRP (Pasha et al. 2016) described the *multi-period FSMVRP* (MPFSMVRP) where the vehicle fleet must service the same set of customers on different days and where the customers' demands vary from day to day. The objective is to find the best possible fleet mix to serve the customers over the full time horizon, and the best possible vehicle routes for each period. As opposed to the *period VRP* (Francis et al. 2008), it is assumed that all customers must be serviced in all periods. (Pasha et al. 2016) show that neither of six simple heuristics for determining the fleet size and mix are able to consistently produce good results, indicating that the resulting problem is more difficult than the FSMVRP.

VRPs containing one or more features that are uncertain during the planning or the implementing phase are known as stochastic VRPs (Toth and Vigo 2002). Typical examples of uncertainty in SVRPs include whether or not a customer requires service, the demand of the customer, or the travel times and service times. Researchers have been studying various stochastic aspects associated with VRPs, but the attention is mostly limited to situations with a given, homogeneous fleet (Kenyon and Morton 2001).

Fleet composition and routing problems have been extensively studied in the last three decades (Hoff et al. 2010; Pantuso et al. 2014). Researchers have mainly focused on deterministic versions of fleet composition problems, and the problems having stochastic features have not gained a similar amount of attention (Hoff et al. 2010). One exception is (Teodorovic et al. 1995), where the demand of each customer is given by a uniform distribution. If the demand on a route exceeds the vehicle capacity a failure occurs, in which case the vehicle has to return to the depot before continuing the route at the next planned customer. The mean total distance of a route, including returns to the depot, is calculated as a function of the route distance. The paper describes a heuristic procedure for creating a giant route which is split into single routes and assigned to appropriate vehicles. The problem shares some similarities with the MPFSMVRPSD, except that it does not cover different periods, and it uses the route distance to calculate the penalties for a route failure. Other examples of stochastic fleet composition problems exist for maritime applications, but contain only limited routing decisions. For example (Gundegjerde et al. 2015) presented a three-stage stochastic programming model for determining the fleet size and mix to support maintenance activities at offshore wind farms. Their model includes variables representing the routing between an onshore depot and the wind farms, but is only accurate for up to two wind farms.

While mostly neglected in the FSMVRP literature, stochastic demands appear in several VRPs with fixed fleets. The VRP with stochastic demands (VRPSD) was first studied by (Bertsimas 1992). The customers' demands are initially only known through a probability distribution. Loaded vehicles with a fixed capacity start at the depot, serve a number of customers on their routes, and return to the depot. Since the demand is a stochastic variable, a vehicle may need to return to the depot if it runs out of stock. A vehicle may return to the depot more than once for refilling if the observed demand exceeds the vehicle capacity, and the route continues until all customers are served. The objective function is the expected total travel length, taking into account return trips to the depot. (Bianchi et al. 2004) showed that the length of the planned route can be used as an approximation of the computationally demanding objective function when using heuristics to solve the VRPSD. This idea was further developed by (Bianchi et al. 2006). Ritzinger et al. outline additional variants of the VRPSD, including references to research where it is treated as a dynamic problem and changes to the vehicle itineraries are allowed after customer visits (Ritzinger et al. 2016).

9.3 Problem Description

The MPFSMVRPSD consists of determining an optimal vehicle fleet size and mix and the corresponding routes to service a set of customers from a single depot. The demands of the customers vary from time period to time period, and separate vehicle routes can be followed in each period, with customers being served by different vehicles in different periods.

Within a given period, the demand of a customer is initially described through a discrete probability distribution and the actual demand becomes known only when the customer is visited. This means that a given route may experience a failure, should the observed total demand from all the customers exceed the vehicle capacity. If, when servicing a customer, the vehicle is emptied, it must return to the depot to refill before recommencing service at the customer. If a vehicle becomes empty while exactly covering the demand of the current customer, it may return to the depot to refill and commence service at the next customer. This may happen since we assume that the demand is discrete. The total expected demand on a route is not allowed to exceed the vehicle capacity. This restriction is included to avoid solutions where a single route serves all the customers (Yang et al. 2000).

To formulate the problem we assume that the probability distributions of the customers' demands have been discretized in the form of scenarios. Pantuso et al. (2015) made a similar assumption for a maritime fleet size and mix problem. Several existing formulations for the VRPSD are based on explicitly using continuous distributions, see for example (Laporte et al. 2002).

Figure 9.1 shows a graphical representation of the MPFSMVRPSD, where the different periods are denoted as days of the week. The main question on level one is which fleet composition to use in order to meet the demand in the best possible way for all periods. This fleet has a fixed cost and remains unchanged throughout the whole week. The customers' demands vary from period to period. Thus, the question on level two is to find appropriate routes for each period using the chosen fleet. The total demand for the customers on any route should not exceed the capacity of the vehicle assigned to that route. The planned routes have a given variable cost if performed without route failures. The execution of the planned routes will, however, depend on the realization of the stochastic demand. This is shown on level three where a given number of scenarios, s, are created for each period. The routes are analyzed for each scenario, and when the vehicle capacity is violated, the cost of a trip back and forth to the depot is incurred. The objective function value equals the sum of the fixed acquisition costs plus the expected costs for following the planned routes (including required refill trips) for each period. A formal definition of the problem is provided below, corresponding to a two-stage stochastic programming problem with recourse.

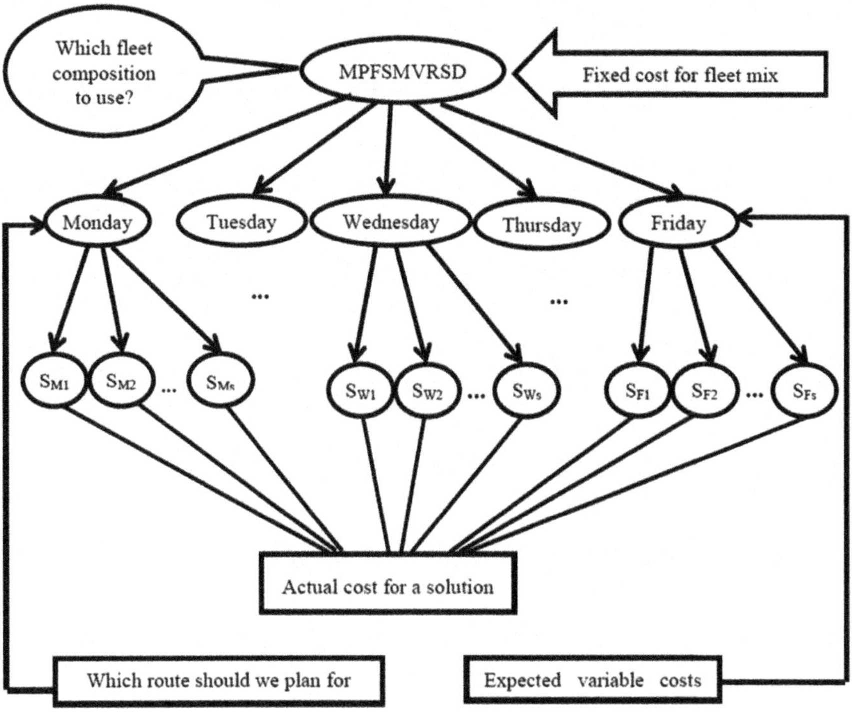

Fig. 9.1 Illustration of the MPFSMVRPSD

Sets:

- T set of time periods
- S_t set of scenarios for time period t
- C set of customers, $C = \{1, \ldots, n\}$, where n is the number of customers
- K set of vehicles types
- N set of all nodes (customers) including the depot which is represented by 0 (start depot) and n+1 (end depot)
- A set of arcs

Parameters:

- c_{kij} cost of traveling from i to j using a vehicle of type k
- g_{ki} cost of requiring a refill when servicing customer i using a vehicle of type k
- h_{kij} additional cost incurred by a return trip to the depot between a visit to node i and node j
- d_{tsj} demand of customer j in time period t, scenario s

p_{ts} probability of scenario s in time period t
q_k capacity of vehicle type k
f_k fixed acquisition cost for a vehicle of type k (per period)

Variables:

x_{tkij} binary decision variable, equal to 1 if a vehicle of type k plans to travel directly from node i to node j in time period t, and 0 otherwise
u_{tski} integer decision variable, equal to the number of times a vehicle of type k needs to return to the depot to refill when servicing customer i in time period t, scenario s
v_{tskij} binary decision variable, equal to 1 if a vehicle of type k refills at the depot between visits at node i and then node j
y_{tskij} flow of goods into node j in time period t, scenario s, using a vehicle that visited node i directly before node j (possibly with a return trip to the depot in between)
e_{ti} expected demand from customers on the route leading up to and including node i in time period t
w_k number of vehicles of type k that are acquired

Formulations:

$$\min Z = \sum_{k\in K} f_k w_k + \sum_{t\in T}\sum_{k\in K}\sum_{(i,j)\in A} c_{kij}x_{tkij}$$
$$+\sum_{t\in T}\sum_{k\in K}\sum_{s\in S_t}\sum_{i\in C} p_{ts}g_{ki}u_{tski} + \sum_{t\in T}\sum_{k\in K}\sum_{s\in S_t}\sum_{i,j\in C} p_{ts}h_{kij}v_{tskij} \tag{9.1}$$

Subject to:

$$\sum_{k\in K}\sum_{(i,j)\in A} x_{tkij} = 1, \qquad t\in T,\ j\in C \tag{9.2}$$

$$\sum_{(i,j)\in A} x_{tkij} - \sum_{(j,i)\in A} x_{tkji} = 0, \qquad t\in T,\ k\in K,\ j\in C \tag{9.3}$$

$$\sum_{(i,j)\in A} y_{tsij} + \sum_{i\in C}\sum_{k\in K} q_k v_{tskij} - \sum_{(j,i)\in A} y_{tsji}$$
$$= d_{tsj} - \sum_{k\in K} q_k u_{tskj}, \qquad t\in T,\ s\in S_t,\ j\in C \tag{9.4}$$

$$y_{ts0j} = \sum_{k \in K} q_k x_{tk0j}, \qquad t \in T,\ s \in S_t,\ (0, j) \in A \tag{9.5}$$

$$y_{tsij} \leq \max_{m \in K} \{q_m\} \left(1 - v_{tskij}\right), \quad t \in T,\ s \in S_t,\ k \in K,\ i \in C,\ j \in C,\ i \neq j \tag{9.6}$$

$$u_{tskj} \leq \sum_{(i,j) \in A} x_{tkij}, \qquad t \in T,\ s \in S_t,\ k \in K,\ j \in C, \tag{9.7}$$

$$v_{tskij} \leq x_{tkij}, \qquad t \in T,\ s \in S_t,\ k \in K,\ i \in C,\ j \in C,\ i \neq j \tag{9.8}$$

$$y_{tsij} \leq \sum_{k \in K} q_k x_{tkij}, \qquad t \in T,\ s \in S_t,\ (i, j) \in A \tag{9.9}$$

$$e_{tj} \geq e_{ti} + \sum_{s \in S_t} p_{ts} d_{tsj} - M_{tij} \left(1 - \sum_{k \in K} x_{tkij}\right), \qquad t \in T,\ (i, j) \in A,\ j \neq n+1 \tag{9.10}$$

$$e_{tj} \leq \sum_{k \in K} q_k \sum_{(i,j) \in A} x_{tkij}, \qquad t \in T,\ j \in C \tag{9.11}$$

$$\sum_{j \in C} x_{tk0j} - w_k \leq 0, \qquad t \in T,\ k \in K \tag{9.12}$$

$$y_{tsij} \geq 0, \qquad t \in T,\ s \in S_t, (i, j) \in A \tag{9.13}$$

$$e_{ti} \geq 0, \qquad t \in T,\ i \in C \cup \{0\} \tag{9.14}$$

$$x_{tkij} \in \{0, 1\}, \qquad t \in T,\ k \in K,\ (i, j) \in A \tag{9.15}$$

$$u_{tski} \geq 0 \text{ and integer}, \qquad t \in T,\ s \in S_t,\ k \in K,\ i \in C \tag{9.16}$$

$$v_{tskij} \in \{0, 1\}, \qquad t \in T,\ s \in S_t,\ k \in K,\ i \in C,\ j \in C,\ i \neq j \tag{9.17}$$

$$w_k \geq 0 \text{ and integer}, \qquad k \in K \tag{9.18}$$

The objective function (9.1) adds up the vehicle fixed acquisition costs, the first stage variable travel costs, and the expected recourse costs. The recourse costs are due to return trips to the depot in order to refill the vehicle. Constraints (9.2) ensure that all customers are visited exactly once in each period. Constraints (9.3) imply that all vehicles which arrive at a customer node must leave that customer node.

The demand of each customer must be fulfilled, as indicated in constraints (9.4). This reduces the current load by a given amount, depending on the scenario, and if needed the vehicle will have to go back to the depot to refill. It is also possible that the vehicle was just refilled before arriving a customer. Constraints (9.5) make sure that the vehicle is full when initially leaving the depot, and constraints (9.6) make sure that the vehicle can only refill between two customer visits if the vehicle is exactly empty when finishing service at the first of the two customers. Only vehicles that visit a customer can refill while servicing that customer, as stated by constraints (9.7), and vehicles can only refill between two customer visits if those are consecutive customers in the planned route, as set by constraints (9.8). The total load on a vehicle may not exceed the vehicle capacity, following constraints (9.9).

Constraints (9.10) make sure that the expected demand on a route up to a given visit is calculated correctly, using

$$M_{tij} = \max_{m \in K} \{q_m\} + \sum_{s \in S_t} p_{ts} d_{tsj}, \tag{9.19}$$

and also prevents the formation of subcycles. The expected demand must be less than the vehicle capacity, as stated by constraints (9.11). Constraints (9.12) make sure that the vehicles used are included in the fleet, and constraints (9.13)–(9.18) specify non-negativity, binary, and integrality constraints for the variables, as needed.

9.4 Solution Method

A good solution to the MPFSMVRPSD minimizes the cost of keeping a vehicle fleet with sufficient capacity, the cost of the routes used to visit the customers, and the cost of additional unplanned returns to the depot for refilling. The heuristic proposed in this chapter first generates an initial solution by solving a deterministic MPFSMVRP and then performs a local search to reduce the costs of routes and unplanned returns to the depot. Finally, a search is conducted to find a better fleet composition, with every change in the fleet composition being followed by the local search to improve routing decisions.

9.4.1 Initial Solution

To create an initial solution, the best strategy found by Pasha et al. 2016 for solving the deterministic MPFSMVRP is used. A deterministic MPFSMVRP is first obtained by taking, for each period, the expected demand for each customer. Then, each period is considered as a separate FSMVRP instance, which is solved heuristically. A fleet composition for the deterministic MPFSMVRP is then selected as the fleet with the largest total capacity amongst the solutions to the separate FSMVRP instances. The vehicle routes are different in each period due to the variable demand, and in some periods, vehicles might be idle as the demand is lower than the capacity of the remaining fleet. The initial solution found in this way, is typically feasible with respect to the expected demand for all periods, but the stochasticity will lead to a number of unplanned returns to the depot on some routes.

9.4.2 *Local Search for Improving Routes*

As the routes in the initial solution are adjusted to meet the requirements of a problem with deterministic demand, they will typically not represent the best routes for the stochastic case. When a change to the fleet mix is performed, the routes need to be adjusted according to the new fleet mix. Thus, to determine the routes in a solution, a simple local search with the chosen fleet is performed. The local search is implemented with a combined neighborhood based on 1-interchange, using either shift moves, which remove one customer from its existing route and insert it to another route, or swap moves, which swap two customers between two different routes. A simple re-optimization procedure, consisting of a local search based on a 2-opt neighborhood, is performed on the affected routes to make sure that the single routes are executed effectively. The local search is repeated as long as the search is improving and terminates when there are no improving solutions in the defined neighborhood.

9.4.3 *Search for Improving the Fleet Composition*

To improve the fleet composition, a new neighborhood involving different moves is defined below. Decisions regarding the fleet change are related to the calculated excess demand and slack in a solution, which again influences the necessary number of returns for refilling. For a given route in a given period for a given scenario, the excess demand is calculated as the maximum of zero and the difference between the total observed demands and the vehicle capacity. Similarly, the slack is the maximum of zero and the difference between the vehicle capacity and the total observed demand.

Before choosing a move, a decision is made regarding whether to increase or decrease the total capacity of the fleet. To find the right balance one needs to determine the level of slack and excess demand that is favorable. Thus, a move that increases the total capacity and improves the objective value from the previous value would indicate that the search may continue in the same direction by increasing capacity. Similarly, an improving move, which decreases capacity, can also be followed by another decreasing move. On the other hand, when a move leads to a worse solution than the previous one, this indicates that the direction of the search should change.

9.4.3.1 Potential Moves for Increasing the Capacity

The neighborhood defines four different ways of changing the fleet when the capacity is increased. Section 9.4.4 presents three different strategies for selecting one of the moves listed.

Add vehicle: One vehicle is added to the fleet. This should always be of the smallest type, as it can be increased in a later iteration if this is considered propitious. If there is overcapacity in a solution, defined as the presence of vehicles serving no customers in all the scenarios used for evaluation, this move cannot be performed.

Increase vehicle size: One of the vehicles in the fleet is exchanged with a vehicle with larger capacity. The strategies considered have different ways of deciding which vehicle to choose. This move cannot be performed if all the vehicles in the fleet are of the largest type.

Merge two small vehicles to a vehicle with a larger total capacity: The two vehicles in the fleet that have the largest average excess demand in all the scenarios used for evaluation, and are small enough to be merged, are chosen. The added vehicle is as small as possible, while having a larger capacity than the two original vehicles taken together. This move cannot be performed if there is no combination of two vehicles in the fleet that can be replaced by one vehicle with a larger total capacity.

Split a vehicle into two smaller vehicles with a larger combined capacity: The vehicle in the fleet which has the largest average excess demand in all the scenarios used for evaluation, is chosen for a split. The added vehicles are such that their combined capacity is as small as possible yet larger than that of the original vehicle. This move cannot be performed if the chosen vehicle is of a type with more than twice the capacity of all the smaller vehicles.

9.4.3.2 Potential Moves for Decreasing the Capacity

The neighborhood defines four different ways of changing the fleet when the capacity is decreased. Section 9.4.4 presents three different strategies for selecting one of the moves listed.

Remove vehicle: One vehicle is removed from the fleet. This should always be of the smallest type included in the fleet, as the others can be decreased in a later iteration if this is considered propitious. If there is insufficient capacity in a solution, that is, all vehicles are in use in all the scenarios used for evaluation, this move cannot be performed.

Decrease vehicle size: One of the vehicles in the fleet is exchanged by a vehicle with a smaller capacity. The strategies considered have different ways of deciding which vehicle to choose. This move cannot be performed if all vehicles in the fleet are of the smallest type.

Merge two vehicles to a vehicle with a smaller total capacity: The two vehicles in the fleet, which have largest slack in all the scenarios used for evaluation, are chosen. The added vehicle is as large as possible, while having a capacity that is smaller than the total capacity of the two original vehicles. This move can always be performed, but could be identical to simply removing a vehicle from the fleet.

Split one vehicle into two vehicles with a smaller combined capacity: The vehicle in the fleet which has the largest slack in all the scenarios used for evaluation, is chosen for a split. The added vehicles are such that their combined capacity is as large as possible yet smaller than that of the original vehicle. This move cannot be performed if the chosen vehicle type has less than twice the capacity of the smallest possible vehicle type.

9.4.3.3 Penalizing Infeasible Solutions

A solution is feasible if the total expected demand of the customers in all routes is smaller than the capacity of the vehicle allocated to that route. Still, during the search, infeasible solutions can appear that needs to be compared with other solutions, both feasible and infeasible. In our search, infeasible solutions are penalized by adding a penalty value to the objective function. The size of the penalty value is found by calculating the positive difference between the total expected demand and the total vehicle capacity for all routes, and multiplying with a large fixed number (1000000 in our tests). In this way, any feasible solution is preferred over an infeasible solution, and the search will quickly return to the feasible space if the fleet's capacity is too small.

9.4.3.4 Preventing Loops in the Search

With the different moves defined in Sects. 9.4.3.1 and 9.4.3.2, the search can easily end up cycling between identical fleet compositions. It is therefore prohibited to perform a move of a type that is the opposite to the type of the previously performed move, to avoid that the fleet mix goes directly back to a combination of vehicles that were explored earlier. If a vehicle is added to the fleet, it should not be removed again in the next iteration, and if the size of a vehicle is increased, it should not be decreased immediately. Similarly, if a split to larger move is performed, a merge to smaller move should not be performed in the next iteration. Thus, no moves that lead back to the fleet from the previous iteration are considered. However, loops can still occur when a previously explored fleet mix is reappearing later in the search. In our search, a loop is detected if three consecutive moves result in the same fleet mix as before the moves were made. If this situation appears several times, a diversification procedure is performed, where all vehicles in the fleet are replaced with a new random vehicle type, and the search continues using this new fleet mix.

1. Find an initial solution by heuristically solving a MPFSMVRP based on the customers' expected demand.
2. Run a local search to improve routes with the current fleet mix.
3. Set the initial direction of the search to Increase capacity.
4. Perform one of the described techniques for selecting a move.
5. Keep track of the last change.
6. Run a local search to improve the routes with the new fleet mix.
7. If the current objective value is better than the previous, continue in same direction, if not change direction.
8. If the objective value is better than the best so far, store the current solution as the best so far.
9. Check whether the current fleet mix has been explored earlier, and if so increase the counter for looping.
10. If the counter for looping is higher than 3, create a new fleet mix by changing the current vehicles to randomly selected vehicle types.
11. Continue from step 4 for a given number of iterations.
12. End the search and report the best found solution.

Fig. 9.2 Main structure of the search

9.4.4 *Overall Search and Three Alternative Techniques*

To search for the best fleet mix, we propose three different techniques to determine the selection of changes to the vehicle fleet. The idea behind two of the techniques is to analyze the current solution and find indications, which should lead to the most favorable fleet change. One possibility is to decide how large the increase or reduction of capacity should be based on the excess demand or slack on the routes. Another possibility is to find the single vehicle with poorest performance on the routes, and change that vehicle. The third possibility is based on making random fleet adjustments. The general search is described in Fig. 9.2. Step 4 states that one of the three techniques for selecting a move should be performed, and the three possible options are described in the next subsections.

9.4.4.1 Average Excess-Slack Technique (AES)

The average excess-slack technique analyzes a solution and tries to determine the preferred change of total capacity. Then the fleet change that best matches the preferred value is performed. The options of adding/removing or increasing/decreasing the size of a vehicle is considered first and the options of merging or splitting are only considered when other constraints in the search makes it necessary to try some other moves. When increasing the capacity the average excess demand is used as the change parameter, while when decreasing capacity, the average slack value divided by the average excess demand is used. The full algorithm is presented in the below:

Algorithm: The average excess-slack technique for deciding fleet change.

1. If the direction of the search is Increase capacity:
   a. Define the parameter **averageExcess** as the average over all periods and scenarios of the total excess for all vehicles.
   b. Find different options for increasing the capacity of one vehicle.
      i. Define the parameter **addCapacity** as the size of the smallest vehicle.
      ii. Define the parameters **increaseCapacity** as all possible options of a larger vehicle type for all of the current vehicles in the fleet.
      iii. Define the parameter **overCapacity** as a Boolean variable that is true if and only if the fleet contains a vehicle that is not used in any of the periods.
   c. If not **overCapacity** and **addCapacity** is closer to average excess demand than any **increaseCapacity** and last change is different from Remove vehicle:
      i. Add a new vehicle of the smallest type.
   d. *Else:*

- *If not all vehicles are of largest size and last change is different from Decrease vehicle size:*
  - *Increase vehicle size for a vehicle that is not of largest type and where* ***increaseCapacity*** *is closest to average excess demand.*
- *Else:*
  - *Choose between Merge to larger and Split to larger.*
    - *If the last change is different from Split to smaller:*
      - *Merge two vehicles to a vehicle of larger total capacity: 1) Choose the vehicles of two routes with largest excess demand that can be merged to a vehicle with larger capacity. 2) Select the new vehicle as the one with closest but larger capacity compared to the sum of two original vehicles.*
    - *If the last change is different from Merge to smaller:*
      - *Split a vehicle into two vehicles of larger total capacity: 1) Choose the vehicle on the route with the largest excess demand that can be split into two vehicles with larger capacity. 2) Select the new vehicles as the two with closest but larger total capacity compared to the original vehicle.*
    - *If there are potential moves from both a and b, the excess demand on the single route selected for splitting is compared with the combined excess demand on the two routes selected for merging. The alternative with the highest excess demand is chosen.*

---

2. Else:

   a. Define the parameter **averageSlackExcess** as the average total slack divided by the average total excess over all periods and scenarios in the current solution, calculating the total slack and excess over all vehicles.
   b. Find different options of reducing the capacity of one vehicle.
      i. Define the parameter **removeCapacity** as the size of the smallest vehicle in the fleet.
      ii. Define the parameters **decreaseCapacity** as all possible options of a smaller vehicle type for each of the current vehicles in the fleet.
      iii. Define the parameter **underCapacity** as a Boolean variable, which is true if and only if all the vehicles in the fleet is utilized in all the periods.
   c. If not **underCapacity** and **removeCapacity** is closer to **averageSlackExcess** than any **decreaseCapacity** and last change is different from Add vehicle:
      i. Remove the smallest vehicle in the fleet.
   d. *Else:*

- *If not all vehicles are of the smallest size and the last change is different from Increase vehicle size:*
  - *Decrease the vehicle size for the vehicle in the fleet that is not of the smallest type and where**decreaseCapacity** is closest to**averageSlack Excess**.*
- *Else*
  - *Choose between Merge to smaller and Split to smaller.*
    - *If the last change is different from Split to larger:*
      - *Merge two routes to a vehicle of smaller total capacity: 1) Choose the vehicles of two routes with the largest slack, which can be merged to a vehicle with smaller capacity. 2) Select the new vehicle as the one with closest but smaller capacity compared to the sum of two original vehicles.*
    - *If the last change is different from Merge to larger*
      - *Split a route into two vehicles with a smaller total capacity. 1) Choose the vehicle on the route with the largest slack, which can be split into two vehicles with a smaller capacity. 2) Select the new vehicles as the two with closest but smaller total capacity compared to the original vehicle.*
    - *If there are potential moves from both a and b, the slack on the single route selected for splitting is compared with the combined slack on the two routes selected for merging. The alternative with the highest slack is chosen.*

9.4.4.2 Single Vehicle Technique (SVT)

The single vehicle technique focuses on the individual vehicle with the worst performance, rather than the fleet in general. The routes for all the vehicles in the fleet are analyzed and, depending on the direction of the search, the slack or the excess demand for the vehicles are compared. The vehicle with the highest value is preferably chosen for an increase or decrease of the size. If this vehicle already is of the largest or smallest vehicle type, or the change is not legal due to other constraints in the search, the other moves in the same direction are considered instead. The full algorithm is presented in the below:

Algorithm: The single vehicle technique for deciding fleet change.

1. If the direction of the search is equal to Increase capacity:
    a. Find the vehicle with the largest excess demand over all periods and scenarios in the current solution.
    b. If this is not of the largest type and last change is different from Decrease vehicle size:
        i. Increase to a larger vehicle type.
    c. Else:
        i. Define the parameter **overCapacity** as a Boolean variable that is true if and only if the fleet contains a vehicle that is not used in any of the periods.
        ii. If not **overCapacity** and the last change is different from Remove vehicle:
            A. Add a new vehicle of the smallest type.
        iii. *Else:*

- *Choose between Merge to larger and Split to larger.*
    - *(a) If the last change is different from Split to smaller:*
        - *Merge two routes to a vehicle of larger total capacity. 1) Choose the vehicles of the two routes with the largest excess demand that can be merged to a vehicle with larger capacity. 2). Select the new vehicle as the one with the closest but larger capacity compared to the sum of two original vehicles.*
    - *(b) If the last change is different from Merge to smaller:*
        - *Split a route into two vehicles of larger total capacity. 1) Choose the vehicle on the route with largest excess demand that can be split into two vehicles with a larger capacity. 2) Select the new vehicles as the two with closest but larger total capacity compared to the original vehicle.*
    - *(c) If there are potential moves from both (a) and (b), the excess demand on the single route selected for splitting is compared with the combined*

excess demand on the two routes selected for merging. The alternative with the highest excess demand is chosen as the move.

2. Else:
   a. Find the vehicle with the largest slack over all periods and scenarios in the current solution.
   b. If this is not of the smallest type and the last change is different from Increase vehicle size:
      i. Reduce to a smaller vehicle type.
   c. Else:
      i. Define the parameter **underCapacity** as a Boolean variable, which is true if and only if all the vehicles in the fleet is utilized in all the periods.
      ii. If not **underCapacity** and last change is different from Add vehicle:
         A. Remove the vehicle from the fleet.
      iii. *Else:*

- *Choose between Merge to smaller and Split to smaller.*
  - *(a) If the last change is different from Split to larger:*
    - *Merge two routes to a vehicle of smaller total capacity. 1) Choose the vehicles of the two routes with the largest slack that can be merged into a vehicle with smaller capacity. 2) Select the new vehicle as the one with the closest but smaller capacity compared to the sum of the two original vehicles.*
  - *(b) If the last change is different from Merge to larger:*
    - *Split a route into two vehicles of smaller total capacity 1) Choose the vehicle on the route with the largest slack that can be split into two vehicles with smaller capacity. 2) Select the new vehicles as the two with the closest but smaller total capacity compared to the original vehicle.*
  - *If there are potential moves from both (a) and (b), the slack on the single route selected for splitting is compared with the combined slack on the two routes selected for merging. The alternative with the highest slack is chosen as the move.*

9.4.4.3 Random Technique (RND)

The third technique is very simple. It is still using the main structure from Fig. 9.2, but instead of analyzing solutions and trying to find indications of what is a good fleet change, a random choice is taken. If the search is going in the increase direction, one of the four possible moves for increasing the total capacity is chosen by random. If the search is going in the decrease direction, one of the four possible moves for decreasing the total capacity is chosen by random. In addition, the vehicles to be involved in the move are also chosen randomly among all possible candidates.

9.5 Computational Study

The heuristics described in Sect. 9.4.4 were implemented using Microsoft Visual Studio C++ and the final compilation of code was done on Intel (R) Xeon (R) machines with 3.4 GHz processors and running the Windows Server 2008 operating system.

9.5.1 Data Sets

Test instances for the MPFSMVRPSD have been generated based on a subset of the instances used by (Pasha et al. 2016) for the MPFSMVRP. These are the Euclidian instances with fixed vehicle acquisition costs, which again were derived from the standard data sets for the FSMVRP defined by (Golden et al. 1984). The characteristics of the instances are shown in Table 9.1.

The expected demand is different in distinct periods, and as in (Pasha et al. 2016), a week with five working days is considered. To include the stochasticity, a percentage

Table 9.1 Overview of data instances used in testing

Instances	Num. of vehicle types	Vehicle costs	Vehicle capacities	Num. of customers
3E	5	20, 35, 50, 120, 225	20, 30, 40, 70, 120	20
4E	3	1000, 1500, 3000	60, 80, 150	20
5E	5	20, 35, 50, 120, 225	20, 30, 40, 70, 120	20
6E	3	1000, 1500, 3000	60, 80, 150	20
13E	6	20, 35, 50, 120, 225, 400	20, 30, 40, 70, 120, 200	50
14E	3	1000, 1500, 3500	120, 160, 300	50
15E	3	100, 250, 450	50, 100, 160	50
16E	3	100, 200, 400	40, 80, 140	50
17E	4	25, 80, 150, 320	50, 120, 200, 350	75
18E	6	10, 35, 100, 180, 400, 800	20, 50, 100, 150, 250, 400	75
19E	3	500, 1200, 2100	100, 200, 300	100
20E	3	100, 300, 500	60, 140, 200	100

Table 9.2 Expected demand per day on different instances

Demands					
Instances	Day 1	Day 2	Day 3	Day 4	Day 5
3E, 4E, 5E, 6E	275	299	374	297	371
13E, 14E	853	917	998	948	961
15E, 16E	647	714	809	761	777
17E, 18E	1187	1380	1366	1275	1309
19E, 20E	1255	1479	1383	1406	1386

deviation from the expected demand is used to define upper and lower bounds for the possible realizations of demand. A fixed number of scenarios where the demands are randomly chosen within the ranges defined by the bounds, are then generated as possible outcomes for each period. The expected total demands on the different days of the planning period are shown in Table 9.2.

Instances based on the data sets by (Golden et al. 1984) have a depot that is located in the middle of the area containing customers, as seen in Fig. 9.3. This is not necessarily a realistic situation: real estate is less costly outside of heavily populated areas and depots and distribution centers are often more likely to be located further away from the customers. Each of the instances described above has therefore also been modified by moving the depot outside of the convex hull defined by the customers. The new instances were defined by looking at the average distance from the depot to all customers in the original instance. Then, the x-coordinate of the depot was moved by adding the average distance to the x-coordinate of the outermost customer. For example in Instance 3E, the depot has coordinates (30, 40) and the average distance from the depot to the customers is 21. The customer with the highest value of the x-coordinate is located on (57, 20), and thus the depot is moved 21 positions from 57, ending up at (78, 40). In the cases where the costumer's coordinates and the vehicle types were identical on two instances, the same technique was performed on the y-coordinate to avoid that the instances became too similar. Figure 9.3 shows the customers and the depot for Instance 3E before and after the depot is moved.

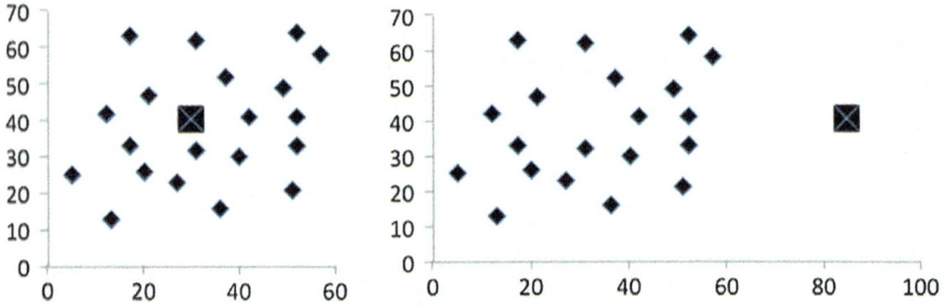

Fig. 9.3 Instance 3E before (*left*) and after (*right*) the depot is moved

We refer to new instances with a moved depot using the same convention as for the original instances, but with "N" appended at the end, instead of "E".

9.5.2 *Effect of the Initial Local Search*

The initial fleet size and mix is determined by solving, heuristically, an expected value problem. The resulting routes may not be ideal for the problem with stochastic demands. Thus, a local search on the given routes is performed initially to check whether the routes can be improved when taking into account the stochastic environment. For these preliminary tests, ten scenarios per day are used. The results show that the local search improves the routing structure with the current fleet composition. Table 9.3 provides comprehensive results by showing the sum of the objective values for all five periods where IS representing the initial solution and IS-LS represent the initial solution after the local search.

The local search is able to improve the routes for all instances, even if the improvement is marginal compared to the total costs. If we compare the routes before and after the local search, we also find some structural differences on the routing. Table 9.4 portrays the details of instance 14E, which is used as a typical example. First, the objective value of the solution before and after the local search is compared. Second, the routing costs for the routes without any returns are shown for the two alternatives. Third, the average number of units in excess on the ten scenarios used for the calculations is shown, and at last, a comparison of the average number of return trips in the ten scenarios is given.

We can see the improvement in the objective value for each of the five periods in the time horizon, and if we only look at the routing costs (including the fixed vehicle costs), without considering the penalties for a return, we can see that they increase slightly after the local search. The average excess demand and the average number of returns is reduced significantly in some of the periods. This indicates that the route utilization has been leveled out, and some demand from routes that originally were almost fully utilized, has now been transferred to routes that had a lower utilization in the deterministic solution. A lower utilization level reduces the risk of an unexpected return. What is even more evident when looking at the routing structure, is that the

Table 9.3 Objective values for initial solutions before and after local search

Instances	3E	4E	5E	6E	13E	14E
IS	5171.7	36766.2	5183.3	36920.5	13044.1	50466.0
IS-LS	5144.5	36752.5	5163.4	36904.4	12945.5	50403.5
Instances	15E	16E	17E	18E	19E	20E
IS	13834.3	14549.0	9304.3	12641.4	48118.0	20952.3
IS-LS	13783.7	14451.0	9203.6	12497.4	48028.1	20804.3

Table 9.4 Routing cost and excess demand for problem instance 14E

Period	Objective value		Routing costs without returns	
	IS	IS-LS	IS	IS-LS
1	10060.4	10040.3	10030.9	10030.5
2	10097.5	10087.6	10058.0	10077.1
3	10112.9	10100.3	10078.8	10078.8
4	10099.0	10085.3	10065.0	10069.8
5	10096.2	10090.0	10069.6	10071.7
Period	Avg. excess demand		Avg. num. of returns	
	IS	IS-LS	IS	IS-LS
1	15.0	5.2	1.7	1.0
2	14.8	4.4	2.5	0.8
3	19.8	19.8	2.1	2.1
4	14.6	10.4	1.5	1.3
5	13.7	11.5	2.2	1.7

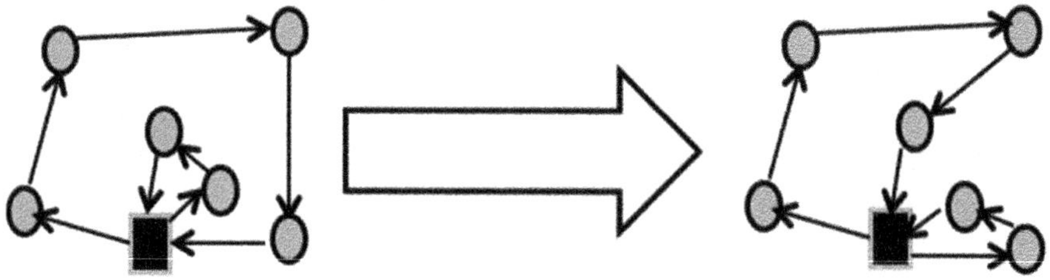

Fig. 9.4 Route structure before and after a local search

customers that are located close to the depot, are now placed either early or late in the different routes. A route violation is more likely to happen at the end of a route, and when an unexpected return to the depot needs to be performed, the cost is smaller when the vehicle is as close as possible. This principle is shown in Fig. 9.4.

9.5.3 Number of Scenarios to Use

When using a model where the probability distributions of the customers' demands have been discretized in the form of scenarios, it is important to determine the number of scenarios to use. To this end, Table 9.5 shows an in-sample stability test where the number of scenarios varies between 1 and 100. The instances are solved using only the initial solution with the local search added. Four of the instances of different sizes were selected for this comparison and 20 runs with different random demands were performed for each combination of instance and number of scenarios.

Table 9.5 Different number of scenarios for initial solutions with a local search

# of scenarios	1	3	5	10	25	50	75	100
Inst. 3								
Average	5101.6	5125.9	5145.7	5151.7	5151.3	5158.1	5152.7	5151.2
Std. dev.	*36.0*	*23.9*	*17.3*	*14.1*	*8.5*	*5.3*	*8.4*	*6.5*
Inst. 13								
Average	12807.8	12871.1	12890.2	12898.4	12913.7	12920.6	12925.2	12926.7
Std. dev.	*44.3*	*29.7*	*27.9*	*24.4*	*11.4*	*15.4*	*15.0*	*12.0*
Inst. 17								
Average	9133.4	9176.1	9194.9	9204.5	9220.2	9216.3	9224.2	9225.8
Std. dev.	*20.2*	*26.7*	*15.3*	*13.6*	*10.9*	*6.8*	*6.0*	*7.0*
Inst. 19								
Average	47986.7	48014.0	48023.4	48033.0	48033.0	48035.7	48036.2	48036.7
Std. dev.	*27.8*	*14.9*	*16.8*	*11.0*	*6.7*	*7.9*	*7.6*	*9.8*

The table shows that the average results are quite stable, and there also seems to be a stabilization of the standard deviation when the number of scenarios reaches 25. Since these tests are based on the initial solution with a local search, we cannot conclude that 25 scenarios is sufficient when considering the full search. Thus, when comparing the different strategies, the total number of scenarios is increased to 50.

9.5.4 Computational Results

Section 9.4.4 describes three techniques for improving the fleet mix in the stochastic problem. One is based on the average excess demand in the solution. Another is analyzing the utilization of each single vehicle to try to find indications for a change, while the third is choosing randomly between the possible options of change. All strategies are based on the major decision of increasing or decreasing the total fleet capacity during the search. A run with 100 iterations for each of the strategies was conducted for the instances with a demand variation of 50% from the expected value and original depot locations. For six of the twelve instances with the depot in its original position, neither of the strategies were able to improve the fleet mix from the initial solution. Table 9.6 shows the results for each of the three techniques relative to the initial solution. In the cases where all strategies give a result of 1.000, the initial solution after the local search is still the best known after the search.

When using different values for the variation in demand, similar results were obtained. Since the final solution is often the same as the initial solution, it is hard to conclude which search technique is preferable. Looking at the results, the AES strategy seems to give the overall best result with an average deviation of 0.9781,

Table 9.6 Relative objective function values for solutions found by different techniques on instances with 50% variance

	3E	4E	5E	6E	13E	14E	
AES	0.9780	0.9440	1.0000	1.0000	0.9891	0.9574	
SVT	1.0000	0.9413	1.0000	1.0000	0.9931	0.9584	
RND	0.9781	1.0000	1.0000	1.0000	0.9833	1.0000	
	15E	16E	17E	18E	19E	20E	Avg.
AES	0.9957	1.0000	1.0000	1.0000	0.9883	1.0000	0.9781
SVT	1.0000	1.0000	1.0000	1.0000	1.0000	1.0000	0.9821
RND	0.9953	1.0000	1.0000	1.0000	0.9979	1.0000	0.9936

compared to SVT with 0.9821, and RND with 0.9936. This indicates that using a technique based on analyzing the solution, is better than a pure random search.

For the instances where the depot has been moved further away from the customer locations, results are clearer. It turns out that with this adjustment, the routes from the initial solution are no longer good options and the search is able to identify both a better distribution of customers between the routes as well as different fleet compositions. Tables 9.7, 9.8 and 9.9 show the relative deviation for the solutions found by the three different techniques, for demands where the maximum deviation is set to 25%, 50%, and 75%, respectively. The value of 1.0000 represents the initial solution on each particular instance. With 25% variation, it seems that SVT gives the best result overall as the average deviation is 0.9324 compared to 0.9352 on AES and 0.9453 on RND. However, with a higher variance, the difference between SVT and AES is levelled out and with 75% variance the difference between the RND strategy and the other strategies is also reduced. One reason could be that the higher variance makes the demand less structured and the advantages of the techniques based on analyzing the solution are reduced.

The overall results of these tests indicate that the solutions of the other two techniques are preferred over the randomized technique. Whether to use the SVT

Table 9.7 Relative objective values for solutions found by different techniques on instances with a maximum deviation in demand of 25%

	3N	4N	5N	6N	13N	14N	
AES	0.8202	1.0000	0.9684	1.0000	0.8643	0.9580	
SVT	0.8174	0.9947	0.9515	0.9993	0.8318	1.0000	
RND	0.8107	1.0000	0.9903	0.9997	0.9102	0.9609	
	15N	16N	17N	18N	19N	20N	Avg.
AES	0.9583	0.9847	0.8912	0.8902	0.9873	1.0000	0.9352
SVT	0.9594	0.9578	0.8369	0.9073	1.0000	0.9863	0.9324
RND	0.9590	0.9594	0.8623	0.8811	1.0000	0.9905	0.9453

Table 9.8 Relative objective values for solutions found by different techniques on instances with a maximum deviation in demand of 50%

	3N	4N	5N	6N	13N	14N	
AES	0.8934	0.9557	0.9545	0.9543	0.8451	0.9594	
SVT	0.8114	0.9946	0.9576	0.9975	0.8237	1.0000	
RND	0.9090	1.0000	0.9547	1.0000	0.9209	1.0000	
	15N	16N	17N	18N	19N	20N	Avg.
AES	0.9597	0.9830	0.8802	0.8577	0.9971	1.0000	0.9270
SVT	0.9550	0.9586	0.8318	0.8697	0.9951	0.9789	0.9308
RND	0.9503	0.9530	0.8616	0.8801	1.0000	0.9803	0.9641

Table 9.9 Relative objective values for solutions found by different techniques on instances with a maximum deviation in demand of 75%

	3N	4N	5N	6N	13N	14N	
AES	0.8071	0.9965	1.0000	0.9564	0.9185	0.9248	
SVT	0.8065	0.9955	0.9522	0.9996	0.8320	1.0000	
RND	0.8071	0.9965	1.0000	0.9564	0.9185	0.9248	
	15N	16N	17N	18N	19N	20N	Avg.
AES	0.9586	0.9691	0.9005	0.8777	0.9976	0.9936	0.9288
SVT	0.9530	0.9491	0.8560	0.8818	1.0000	0.9853	0.9310
RND	0.9500	0.9508	0.8304	0.8868	1.0000	0.9895	0.9339

technique, which analyzes each route, or the AES technique, which rather looks at the total excess demand or slack, is not so obvious.

In Table 9.10, a comparison of the total fleet capacity for the best solutions found with different demand variations is given. Surprisingly, the variance does not seem to affect the necessary capacity significantly, and we can even see that the fleet capacity is reduced slightly on many instances when uncertainty is introduced. The variation can go in both directions, and in general the increased demand for some customers is cancelled out by a decreased demand for others.

This is further analyzed in Table 9.11, which shows the objective function values for the best solutions found for all the instances with relocated depots, under different demand variation. In the cases where the same fleet mix is found for the instances with different demand, a higher variance leads to more excess demand and additional returns to the depot. That is, higher demand variation is associated with a higher total cost. This is seen for example on instance 3 N where the fleet mix consists of four vehicles of the largest type with capacity 120 and one vehicle of the smallest type with capacity 20. However, there are some examples where the cost of the best found solution is lower when the demand deviation is higher. There are two possible explanations. Either the scenarios are, by chance, more balanced for the case with higher demand deviations, such that better fleet utilization is possible, or the search

Table 9.10 Best found fleet capacity with different variation in customer's demand

Fleet capacity				
Instances	IS	25% deviation	50% deviation	75% deviation
3N	390	380	380	380
4N	390	390	380	390
5N	390	380	390	380
6N	390	390	380	380
13N	1030	1020	1000	1020
14N	1040	1000	1000	1000
15N	850	820	840	830
16N	840	820	820	820
17N	1420	1400	1400	1400
18N	1420	1400	1390	1390
19N	1500	1500	1500	1500
20N	1500	1500	1480	1500
Average relative difference		0.9844	0.9809	0.9827

Table 9.11 Objective function values for the best solution found with different demand variance

	3N	4N	5N	6N	13N	14N	
Deviation 25%	6142.9	38106.6	6239.4	38392.5	14379.0	50422.9	
Deviation 50%	6161.3	38104.5	6342.2	36776.7	14360.4	50600.2	
Deviation 75%	6202.5	38203.7	6293.6	36877.5	14463.2	48741.7	
	15N	16N	17N	18N	19N	20N	Avg.
Deviation 25%	16049.9	16649.6	10866.4	16683.6	50808.6	25498.8	0.9270
Deviation 50%	16186.4	16749.3	10953.6	16516.4	51504.0	25702.8	0.9308
Deviation 75%	16249.0	16770.2	10982.6	17026.9	51681.5	25919.6	0.9641

has missed some good solutions for the case with lower demand deviations. An example of an instance where a higher demand deviation gives a lower objective function value for the best solution found is 18N. The fleet mix on 25% deviation has a total capacity of 1400 and consists of six vehicles with capacity 100, two with capacity 150 and two with capacity. For a deviation of 50% the fleet capacity is 1390, consisting of two vehicles with capacity 20, one with capacity 100, five with capacity 150, and finally two with capacity 250. Even if the total capacity is smaller, the new fleet mix seems better fit to deal with the actual scenarios.

9.6 Concluding Remarks

This chapter presents a stochastic problem called the multi-period fleet size and mix vehicle routing problem with stochastic demands (MPFSMVRPSD) where the customers are known but their demand is uncertain. The objective is to find the fleet composition and fixed routing for all periods in the time horizon, which minimizes the expected total transportation cost. The customer demand is not known in advance and a route violation will result in an unplanned return to the depot, which requires an extra cost. An initial solution is constructed based on a technique developed for the deterministic MPFSMVRP by using the expected customer's demand as a fixed value. It is shown that the initial solution can be improved by performing a simple local search, as it is observed that the vehicles usually get empty and have to return to the depot for refilling while serving the last customers on a route. Thus, the routes should be constructed such that the customers located close to the depot also are the last customers on the routes.

The MPFSMVRPSD is a complex and NP-hard problem, which is difficult to solve. In this chapter we proposed three simple techniques for searching for a more suitable fleet. These are based on one major decision, i.e., whether to increase or decrease the total fleet capacity. To compare these techniques, a set of new instances was created based on standard FSMVRP instances. For some of the new instances, the depot is moved further away from the customer locations, which increases the penalty for unplanned returns. Two of the proposed techniques, which make decisions based on analyzing the solution, show better results than a technique picking random moves among those leading in the right direction. The results do not give any clear conclusion whether analyzing single routes or the total excess and slack on the routes is the best strategy for making fleet change decisions. The variance of the customers' demands affects the best found selection of vehicles for the fleet in some instances.

References

Bertsimas DJ (1992) A vehicle routing problem with stochastic demand. Oper Res 40(3):574–585

Bianchi L, Birattari M, Chiarandini M, Manfrin M, Mastrolilli M, Paquete L, Rossi-Doria O, Schiavinotto T (2004) Metaheuristics for the vehicle routing problem with stochastic demands. In: International conference on parallel problem solving from nature. Springer, pp 450–460

Bianchi L, Birattari M, Chiarandini M, Manfrin M, Mastrolilli Monaldo, Paquete Luis, Rossi-Doria Olivia, Schiavinotto Tommaso (2006) Hybrid metaheuristics for the vehicle routing problem with stochastic demands. J Math Modell Alg 5(1):91–110

Dror M, Laporte G, Trudeau P (1989) Vehicle routing with stochastic demands: Properties and solution frameworks. Transp Sci 23(3):166–176

Francis PM, Smilowitz KR, Tzur M (2008) The period vehicle routing problem and its extensions. In: The vehicle routing problem: latest advances and new challenges. Springer, pp 73–102

Golden B, Assad A, Levy L, Gheysens F (1984) The fleet size and mix vehicle routing problem. Comput Oper Res 11(1):49–66

Gundegjerde C, Halvorsen IB, Halvorsen-Weare EE, Hvattum LM, Nonås Lars Magne (2015) A stochastic fleet size and mix model for maintenance operations at offshore wind farms. Transp Res Part C: Emerg Technol 52:74–92

Hoff A, Andersson H, Christiansen M, Hasle G, Løkketangen Arne (2010) Industrial aspects and literature survey: fleet composition and routing. Comput Oper Res 37(12):2041–2061

Kenyon AS, Morton DP (2001) A survey on stochastic location and routing problems. Central Eur J Oper Res 9:277–328

Laporte G, Louveaux FV, Hamme Van L (2002) An integer L-shaped algorithm for the capacitated vehicle routing problem with stochastic demands. Oper Res 50(3):415–423

Pantuso G, Fagerholt K, Hvattum LM (2014) A survey on maritime fleet size and mix problems. Eur J Oper Res 235(2):341–349

Pantuso G, Fagerholt K, Wallace SW (2015) Uncertainty in fleet renewal: a case from maritime transportation. Transp Sci 50(2):390–407

Pasha U, Hoff A, Hvattum LM (2016) Simple heuristics for the multi-period fleet size and mix vehicle routing problem. Inf Syst Oper Res INFOR:1–24

Ritzinger U, Puchinger J, Hartl RF (2016) A survey on dynamic and stochastic vehicle routing problems. Int J Prod Res 54(1):215–231

Teodorovic D, Krcmar-Nozic E, Pavkovic G (1995) The mixed fleet stochastic vehicle routing problem. Transp Plann Technol 19(1):31–43

Toth P, Vigo D (2002) The vehicle routing problem. Society for Industrial and Applied Mathematics. SIAM monographs on discrete mathematics and applications

Yang W-H, Mathur K, Ballou RH (2000) Stochastic vehicle routing problem with restocking. Transp Sci 34(1):99–112

Chapter 10
Applying Multi-objective Robust Design Optimization Procedure to the Route Planning of a Commercial Aircraft

Jordi Pons-Prats, Gabriel Bugeda, Francisco Zarate, Eugenio Oñate and Jacques Periaux

Abstract Aircraft emission targets worldwide and their climatic effects have put pressure in government agencies, aircraft manufacturers and airlines to reduce water vapour, carbon dioxide (CO_2) and oxides of nitrogen (NO_x) resulting from aircraft emissions. The difficulty of reducing emissions including water vapor, carbon dioxide (CO_2) and oxides of nitrogen (NO_x) is mainly due to the fact that a commercial aircraft is usually designed for a particular optimal cruise altitude but may be requested or required to operate and deviate at different altitudes and speeds to archive a desired or commanded flight plan, resulting in increased emissions. This is a multi- disciplinary problem with multiple trade-offs such as optimizing engine efficiency, minimizing fuel burnt and emissions while maintaining prescribed aircraft trajectories, altitude profiles and air safety. There are possible attempts to solve such problems by designing new wing/aircraft shape, new efficient engine, ATM technology, or modifying the aircraft flight plan. Based on the rough data provided by an air carrier company, who was willing to assess the methodology, this paper will present the coupling of an advanced optimization technique with mathematical models and algorithms for aircraft emission, and fuel burnt reduction through flight plan optimization. Two different approaches are presented; the first one describes a deterministic optimization of the flight plan and altitude profile in order to reduce the

J. Pons-Prats (✉) · G. Bugeda · F. Zarate · E. Oñate · J. Periaux
CIMNE, Barcelona, c/Jordi Girona s/n Campus Nord UPC, Edifici C1,
08034 Barcelona, Spain
e-mail: jpons@cimne.upc.edu

G. Bugeda
e-mail: bugeda@cimne.upc.edu

F. Zarate
e-mail: zarate@cimne.upc.edu

E. Oñate
e-mail: onate@cimne.upc.edu

J. Periaux
e-mail: jperiaux@gmail.com

G. Bugeda · E. Oñate
Universitat Politecnica de Catalunya - BarcelonaTech, Barcelona, Spain

P. Diez et al. (eds.), *Computational Methods and Models for Transport*,
Computational Methods in Applied Sciences 45,
DOI 10.1007/978-3-319-54490-8_10

fuel consumption while reducing time and distance. The second approach presents the robust design optimization of the previous case considering uncertainties on several parameters. Numerical results will show that the methods are able to capture a set of useful trade-offs solutions between aircraft range and fuel consumption, as well as fuel consumption and flight time.

10.1 Introduction

Environmental concerns but also the economic issues arising from the operation motivates low-budget air carrier to optimize the route planning in regards fuel consumption, time and distance. Fuel consumption is directly related to the level of emissions, CO_2, NO_x and tiny particles. Time and distance are directly related to the operation and the satisfaction of the customer. The present paper describes a simple test case aimed to demonstrate the capabilities of numerical optimization in order to define the optimum flight procedures and to reduce the fuel consumption. Mission path planning has been investigated in Lee et al. (2012), aiming on Unnamed air vehicles (UAV) and a fuel reduction as the objective functions of an optimization problem, or in Gonzalez et al. (2011), aiming emission reduction as one of the main objective functions of the optimization problem.

The presented test case is based on the real operation, but applying several simplifications due to a non-disclosure agreement with the air carrier involved. A rough model of the fuel consumption has been defined, which adds information to the time and distance calculation onto a specific route. Due to the simplifications, the test case is aimed to validate the methodologies and present a very preliminary results.

The methodology analyses the mission profile, which is split in several phases. Flying time, fuel consumption and distance have been calculated as part of the phase definition. Parameters like the climb ratio and the fuel flow of the engines will define the fuel consumption, all of them will be described later on. A limited number of waypoints have been defined to calculate the distance. Flying velocity is fixed at each phase, so the time has been defined by the ratio of distance and velocity. For comparison purposes, three procedures have been used to solve the problem; namely a deterministic, a stochastic and a robust design procedures. The use of statistical information, as mean and standard deviation of the quantities of interest on the robust, or only the mean of the quantities of interest on the stochastic procedure differentiates one from another, while the deterministic one does not consider any uncertainty at all, so it works directly with the values of the quantity of interest.

The optimization analysis is based on NSGA-II algorithm. NSGA-II is a well-known algorithm, which has been validated and tested in several problems, see (Deb et al. 2000; Deb 2001; Deb et al. 2002a, b) for additional reference. It works well with constrained and unconstrained problems as described in Deb et al. (2001).

10.2 Definition of the Problem

The proposed problem definition comes from the contribution of a low-budget air carrier company. The aim of that discussions were to establish the framework of a research program. The mission planning problem is aimed to the optimization of a flying route between two fixed points. Way-points are defined as intermediate passing through points, which are used as restrictions of the route definition. Also restrictions to the parameters that define the vertical profile of the mission are applied. Climb rate, acceleration and cruise altitudes, restricted within feasible range of values, according to the aircraft performances and air traffic restrictions, are examples of examples of the parameters used on the definition of the problem. Guidelines are taken from (AIRBUS 2008) and (BOEING 2007).

The analysis considers three objective functions. The fuel consumption, the time of the mission and the length of the mission are the three of them. The mission has been divided into phases; taxi, take-off, initial ascent, acceleration at acceleration altitude, ascent to cruise altitude, cruise at first cruise altitude, ascent/descent to second cruise altitude, cruise at second cruise altitude, descent, descent in final approach, and finally landing, as described in Fig. 10.1.

The calculation of the distance between way-points helps to obtain the model for the time and the distance. The velocity is used to calculate the time, but considering the flying path which could slightly modify the real distance and time due to the ascent or descent of the aircraft. The fuel consumption model is an approximation from the data available regarding the selected route and company data. As an approximation, fuel consumption uses a limited set of parameters of the real ones.

Several simplification have been applied to reduce the total amount of variables of the problem. Limitation on the number of flight phases, as well as considering some

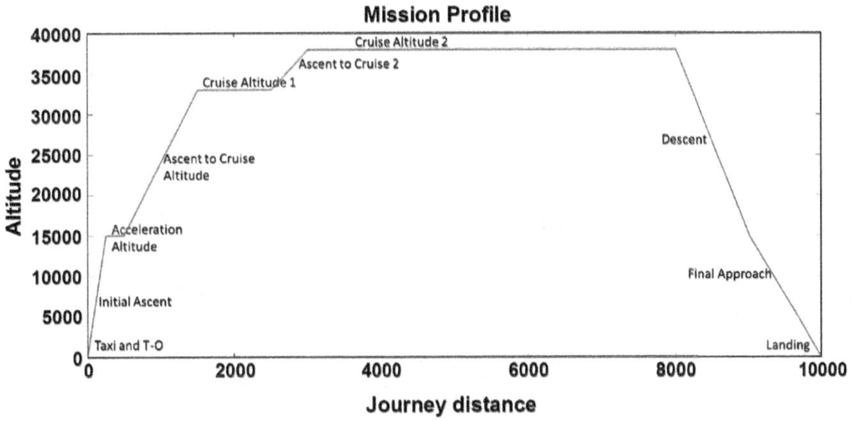

Fig. 10.1 Mission phase

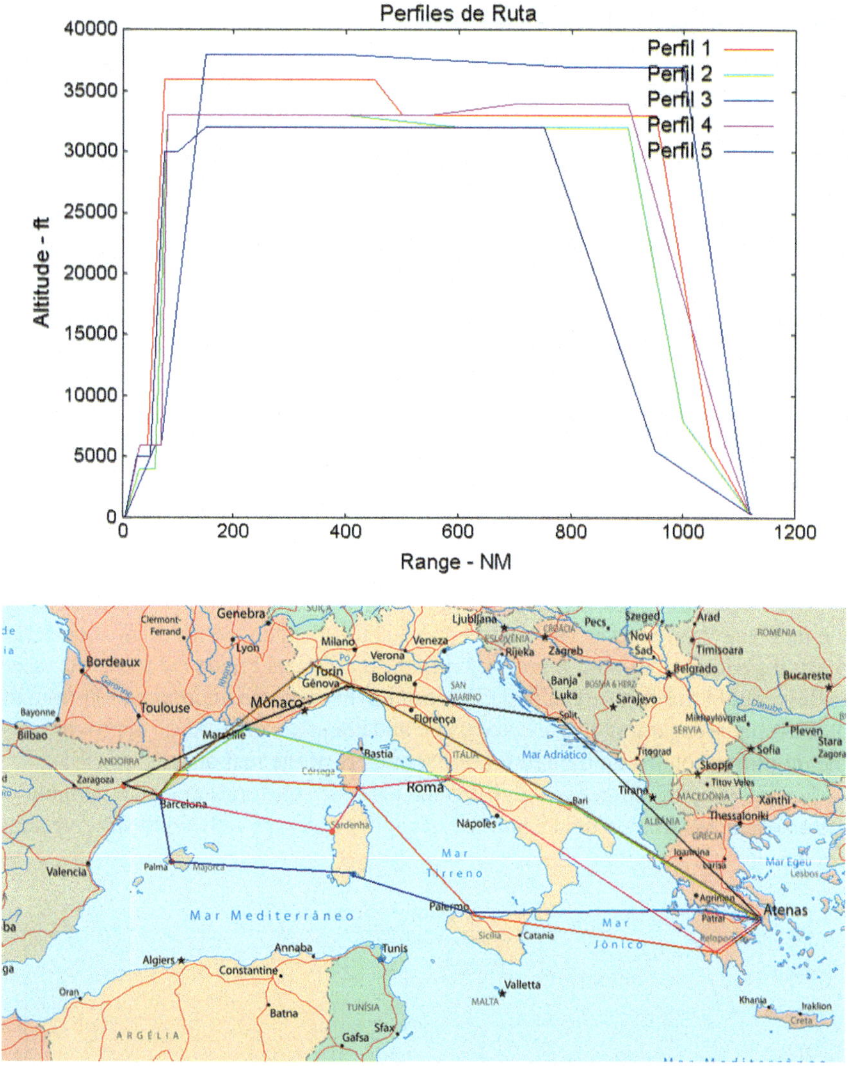

Fig. 10.2 Mission profiles and routes

variables as constant (throttle during cruise and leveled flight phases, Mach number for both first and second cruise are some examples).

The selected route where this analysis applies is from Barcelona, as departure airport, to Athens, as destination. The way-points are limited to 4. Each way-point can be located to several geographical points:

- Departure Point, Barcelona.
- Waypoint 1: Girona, Lleida, Palma de Mallorca, Alghero.

- Waypoint 2: Marseille, Cagliari, Tunis, Porto-Vecchio.
- Waypoint 3: Torino, Genova, Roma, Palermo.
- Waypoint 4: Split, Bari, Patra, Kalamata.
- Destination Point: Athens.

Possible mission profiles and routes are given in Fig. 10.2.

10.3 Defining the Governing Equations

The three objective functions of the optimization problem are defined as the total fuel consumption, FC, the total time, T, and the flown distance, D. The three of them are defined as the sum of the partial values of each phase of the flight:

$$
\begin{aligned}
FC &= fc_{tx} + fc_{to} + fc_{a2aa} + fc_{aa} + fc_{aa2c} + fc_{c1} + \\
&\quad fc_{c2c} + fc_{c2} + fc_{d2a} + fc_{fa} \\
T &= t_{tx} + t_{to} + t_{a2aa} + t_{aa} + t_{aa2c} + t_{c1} + t_{c2c} + \\
&\quad t_{c2} + t_{d2a} + t_{fa} \\
D &= d_{tx} + d_{to} + d_{a2aa} + d_{aa} + d_{aa2c} + d_{c1} + \\
&\quad d_{c2c} + d_{c2} + d_{d2a} + d_{fa}
\end{aligned}
\tag{10.1}
$$

The definition, phase by phase, of the functions is described below. Taxi, the first phase, considers the time value (t_{tx}) as fixed at a value equal to 5 min. The distance is not considered in the overall journey. Fuel consumption during Taxi phase, fc_{tx} is computed as:

$$fc_{tx} = (1 + SE) \cdot ffr \cdot (1 + th_x) \cdot t_{tx} \tag{10.2}$$

t_{tx}: duration of taxi, fixed to 5 min.
ffr: fuel flow reference; minimum fuel flow of each engine (kg/h).
SE: single engine taxi; 0 = taxi with only one engine, 1 = taxi with two engines.
th_{tx}: throttle adjustment (% of full throttle).

Time, t_{to}, and Fuel consumption, fc_{to}, of Take-off phase are computed as:

$$
\begin{aligned}
t_{to} &= \frac{2l_r}{1852 \cdot v_{to}} \\
fc_{to} &= (2ffr - h_{to} + w_{to} \cdot 10^{-3}) \cdot 2t_{to}
\end{aligned}
\tag{10.3}
$$

v_{to}: take-off speed (kt).
l_r: runway length (m).

h_{to}: altitude of departure airfield (ft).
w_{to}: wind during take-off (kt).

Time, t_{a2aa}, distance, d_{a2aa}, and Fuel consumption, fc_{a2aa}, of Ascent phase after take off are computed as:

$$t_{a2aa} = \frac{AA - h_{to}}{CR2AA \cdot 60}$$

$$d_{a2aa} = v_{to} \cdot t_{a2aa} \cdot \cos\left(a \cdot \sin\left(\frac{\frac{AA-h_{to}}{6076.115}}{v_{to} \cdot t_{a2aa}}\right)\right) \tag{10.4}$$

$$fc_{a2aa} = \left(\frac{2ffr - AA \cdot 10^{-3} + w_{to} \cdot (1 + AA \cdot 10^{-3}) \cdot 10^{-3} + AA}{609 \cdot \sqrt{1 + \frac{15 + 1.98 \cdot AA \cdot 10^{-3}}{273.15}}}\right) \cdot 2t_{a2aa}$$

AA: Acceleration altitude (ft).

CR2AA: climb ratio to acceleration altitude (ft/min).

Time, t_{aa}, distance, d_{aa}, and Fuel consumption, fc_{aa}, of flight at acceleration altitude are computed as:

$$t_{aa} = \frac{v_{aa} \cdot 609 \cdot \sqrt{1 + \frac{15 + 1.98 \cdot AA \cdot 10^{-3}}{273.15}} - v_{to}}{12596 \cdot (1 - CI_{aa}) \cdot 10^{-2}}$$

$$d_{aa} = v_{to} \cdot t_{aa} + 0.5 \cdot \left(12596 \cdot (1 - CI_{aa}) \cdot 10^{-2}\right) \cdot t_{aa}^2 \tag{10.5}$$

$$fc_{aa} = \left(\frac{2 \cdot ffr - AA \cdot 10^{-3} + w_{to} \cdot \left(1 + AA \cdot 10^{-3}\right) \cdot 10^{-3} + v_{to}}{609 \cdot \sqrt{1 + \frac{15 + 1.98 \cdot AA \cdot 10^{-3}}{273.15}}}\right) \cdot 5\,(1 + th_{aa}) \cdot t_{aa}$$

v_{aa}: Mach number at acceleration altitude (% Mach).
CI_{aa}: Cost Index during flight to acceleration altitude.
th_{aa}: throttle position during flight at acceleration altitude.

Time, t_{aa2c}, distance, d_{aa2c}, and Fuel consumption, fc_{aa2c}, of ascent to cruise altitude are computed as:

$$t_{aa2c} = \frac{CA_1 - AA}{CR2C \cdot 60}$$

$$d_{aa2c} = v_{aa} \cdot 609 \cdot \sqrt{1 + \frac{15 + 1.98 \cdot CA_1 \cdot 10^{-3}}{273.15}} \cdot t_{aa2c} \cdot \cos\left(a \cdot \sin\left(\frac{\frac{CA_1 - AA}{6076.115}}{v_{aa} \cdot 609 \cdot \sqrt{1 + \frac{15 + 1.98 \cdot CA_1 \cdot 10^{-3}}{273.15}} \cdot t_{aa2c}}\right)\right) \tag{10.6}$$

$$fc_{aa2c} = \left(2ffr - CA_1 \cdot 10^{-3} + w_{c1} \cdot 10^{-3} + v_{c1}\right)(1 + th_{aa})\, t_{aa2c}$$

CR2C: climb ratio to cruise altitude (ft/min).
CA_1: Cruise altitude 1 (ft).
w_{c1}: wind at cruise level 1 (kt).

Time, t_{c1}, distance, d_{c1}, and Fuel consumption, fc_{c1}, of first cruise phase are computed as:

$$\begin{aligned} t_{c1} &= \frac{d_{c1}}{v_{c1} \cdot 609 \cdot \sqrt{1 + \frac{15 + 1.98 \cdot CA_1 \cdot 10^{-3}}{273.15}}} \\ d_{c1} &= \frac{d_{od} - (d_{a2aa} + d_{aa} + d_{aa2c} + d_{c2a} + d_{fa})}{2} \\ fc_{c1} &= \left(\left(2ffr - CA_1 \cdot 10^{-3} + w_{c1} \cdot 10^{-3} + 5v_{c1}\right)(1 + th_{aa}) + \right. \\ &\quad \left. + (100 - CI_{c1} \cdot 2ffr)\right) \cdot t_{c1} \end{aligned} \tag{10.7}$$

v_{c1}: Mach number at cruise level 1.
d_{od}: distance from departure to destination airfields (NM).
CI_{c1}: cost index during cruise 1.

Time, t_{c2c}, distance, d_{c2c}, and Fuel consumption, fc_{c2c}, of ascent/descent from cruise altitude 1 to cruise altitude 2 are computed as:

$$\begin{aligned} t_{c2c} &= \frac{CA_1 - CA_2}{CRCC \cdot 60} \\ d_{c2c} &= v_{c1} \cdot 609 \cdot \sqrt{1 + \frac{15 + 1.98 \cdot CA_2 \cdot 10^{-3}}{273.15}} \cdot t_{c2c} \\ &\quad \cdot \cos\left(a \cdot \sin\left(\frac{\frac{CA_1 - CA_2}{6076.115}}{v_{c1} \cdot 609 \cdot \sqrt{1 + \frac{15 + 1.98 \cdot CA_2 \cdot 10^{-3}}{273.15}} \cdot t_{c2c}}\right)\right) \\ fc_{c2c} &= \left(2ffr - CA_2 \cdot 10^{-3} + w_{c2} \cdot 10^{-3} + 5v_{c1}\right)(1 + th_{aa})\, t_{c2c} \end{aligned} \tag{10.8}$$

CRCC: climb ratio to cruise altitude 2 (ft/min).
CA_2: Cruise altitude 2 (ft).
w_{c2}: wind at cruise level 2 (kt).

Time, t_{c2}, distance, d_{c2}, and Fuel consumption, fc_{c2}, of second cruise phase are computed as:

$$\begin{aligned} t_{c2} &= \frac{d_{c2}}{v_{c1} \cdot 609 \cdot \sqrt{1 + \frac{15 + 1.98 \cdot CA_2 \cdot 10^{-3}}{273.15}}} \\ d_{c2} &= \frac{d_{od} - \left(d_{a2aa} + d_{aa} + d_{aa2c} + d_{c2c} + d_{c2a} + d_{fa}\right)}{2} \end{aligned} \tag{10.9}$$

$$fc_{c2} = \left(\left(2ffr - CA_2 \cdot 10^{-3} + w_{c2} \cdot 10^{-3} + 5v_{c1}\right)(1 + th_{aa}) + (100 - CI_{c2} \cdot 2ffr)\right) \cdot t_{c2}$$

CI_{c2}: cost index during cruise 2.

Time, t_{d2a}, distance, d_{d2a}, and Fuel consumption, fc_{d2a}, of descent phase to final approach altitude are computed as:

$$\begin{aligned} t_{d2a} &= \frac{CA_2-(h_d+3000)}{CRFA\cdot 60} \\ d_{d2a} &= v_{c1} \cdot 609 \cdot \sqrt{1 + \frac{15+1.98\cdot(h_d+3000)\cdot 10^{-3}}{273.15}} \cdot t_{d2c} \\ &\quad \cdot \cos\left(a \cdot \sin\left(\frac{\frac{CA_2-(h_d+3000)}{6076.115}}{v_{c1}\cdot 609\cdot\sqrt{1+\frac{15+1.98\cdot(h_d+3000)\cdot 10^{-3}}{273.15}}\cdot t_{d2c}}\right)\right) \\ fc_{d2a} &= \left(2ffr - CA_2 \cdot 10^{-3} + w_{c2} \cdot 10^{-3} + 5v_{c1}\right)(1 + th_{aa})\, t_{d2a} \end{aligned} \tag{10.10}$$

CRFA: descent ratio to final approach altitude (ft/min) h_d: altitude of the destination airfield (ft)

Time, t_{fa}, distance, d_{fa}, and Fuel consumption, fc_{fa}, of final approach phase are computed as:

$$\begin{aligned} t_{fa} &= 0.6 \\ d_{fa} &= 25 \\ fc_{fa} &= 2ffr \cdot (1 + th_{tx}) \cdot t_{fa} \end{aligned} \tag{10.11}$$

Optimization parameters are defined as follows:

- Throttle taxi (%): [0.15: 0.3]
- Take off speed (kt): [220: 260]
- Acceleration Altitude, AA (ft): [3000: 6000]
- Cost Index at Acceleration Altitude: [0: 50]
- Acceleration Altitude Mach: [0.65: 0.82]
- Throttle at Acceleration Altitude: [0.5: 0.8]
- Climb rate post-AA (ft/min): [500: 1500]
- Cruise alt1 (ft): [30000: 42000]
- Cruise 1 Mach (kt): [0.75: 0.8]
- Climb rate cruise (ft/min): [300: 1000]
- Cruise alt2 (ft): [30000: 42000]
- Descent rate(ft/min): [300: 1000]
- Cost Index Cruise 1: [0: 100]
- Cost Index Cruise 2: [0: 100]
- Single engine taxi: Yes/No, equivalent to values 1 or 0.

Considering constant values:

- Reference Fuel Flow; ffr = 1000 kg/h C
- limb rate to Acceleration altitude; CR2AA = 1000 ft/min
- Runway length for take off; lr = 2500 m
- Wind at Take off: wto = 15 kt
- Wind at Cruise level 1: wc1 = 60 kt
- Wind at Cruise level 2: wc2 = 40 kt

10.4 Deterministic Optimization Procedure

The multi-objective minimization of the functions in equations (10.12) defines the deterministic optimization::

$$
\begin{aligned}
FC_d &= fc_{tx} + fc_{to} + fc_{a2aa} + fc_{aa} + fc_{aa2c} + \\
&\quad fc_{c1} + fc_{c2c} + fc_{c2} + fc_{d2a} + fc_{fa} \\
T_d &= t_{tx} + t_{to} + t_{a2aa} + t_{aa} + t_{aa2c} + t_{c1} + t_{c2c} + \\
&\quad t_{c2} + t_{d2a} + t_{fa} \\
D_d &= d_{tx} + d_{to} + d_{a2aa} + d_{aa} + d_{aa2c} + d_{c1} + \\
&\quad d_{c2c} + d_{c2} + d_{d2a} + d_{fa}
\end{aligned}
\tag{10.12}
$$

No uncertainty in the parameters is considered in the deterministic optimization. Then, the objective functions are directly the functions to calculate the quantity of interest (the fuel consumption, the time and the distance). The NSGA-II optimizer has been set up as:

- Population size: 100
- Number of populations: 5000
- Probability of Crossover: 0,90
- Probability of Mutation: 0,06

10.5 Stochastic and Robust Design Optimization Procedure

The stochastic and robust optimization problems are based on what has been described in Sect. 10.2. In these two cases uncertainty are applied to the parameters. Section 10.5.1 contains the description of the application of the uncertainties to the governing equations. Standard deviation of the quantities of interest has not been defined as an objective function in the stochastic analysis, while it is used in the robust design optimization. The mean of the quantity of interest is the main objective function in both the stochastic and robust analysis.

10.5.1 Applying Uncertainty to the Governing Equations

The stochastic and robust optimization analysis already defines uncertainty in the parameters. These parameters are defined by means of their probability density function, PDF (mainly considering the mean and the standard deviation values). Other kind of uncertain definition adds an uncertain term to the value, also defined by a PDF.

Based on the information provided by a real operator, so with considering real values, the definition of the uncertainty on the parameters tries to approximate the unknown behavior of specific conditions during the mission, like wind direction, temperature at altitude affecting the engine performance, or similar. The cruise speed and cruise altitude are affected by air traffic restrictions, but also the weather conditions can change along the flight adding uncertainty. Those parameters directly modify the calculation of the fuel consumption, the mission time and the distance. Each phase would be affected in a different way, but similarly how it is described in the deterministic case.

Two cases have been considered; namely the stochastic and the robust cases. The stochastic optimization problem considers 3 objective functions. They are the mean values of fuel consumption, the mean of the total time, and the mean of the total distance of the flight. Compared to the deterministic case, the two cases are quite similar. The difference remains on the fact that some parameters are perturbed. The second case is the robust optimization problem. It considers 5 objective functions. These functions are the three mean values, as described in the stochastic case, and the two standard deviation values of fuel and time.

The basic formula to be used along the analysis are modified as described below. Only those equations affected by the definition of uncertainties are listed. Those equations remaining the same can be found in the previous sections. For the Taxi phase, the fuel consumption, fc_{tx}, is:

$$fc_{to} = \left(2\xi_{ffr} - h_{to} \cdot 10^{-3} + \xi_{w_{to}} \cdot 10^{-3}\right) \cdot 2t_{to} \tag{10.13}$$

ξ_{ffr}: uncertain fuel flow
$\xi_{w_{to}}$: uncertain wind during take-off (kt)

For the ascent phase after take-off, the time, t_{a2aa}, and distance, d_{a2aa}, are computed as in the deterministic case, and Fuel consumption, fc_{a2aa}, is:

$$fc_{a2aa} = \left(\frac{2\xi_{ffr} - AA \cdot 10^{-3} + \xi_{w_{to}} \left(1 + AA \cdot 10^{-3}\right) \cdot 10^{-3} + AA}{609 \cdot \sqrt{1 + \frac{15 + 1.98 \cdot AA \cdot 10^{-3}}{273.15}}}\right) \cdot 2t_{a2aa} \tag{10.14}$$

For the phase of flight at acceleration altitude, the time, t_{aa}, and the distance, d_{aa}, are computed as in the deterministic case and the fuel consumption, fc_{aa} is:

$$fc_{aa} = \left(\frac{2\xi_{ffr} - AA\cdot 10^{-3} + \xi_{w_{to}}\cdot\left(1 + AA\cdot 10^{-3}\right)\cdot 10^{-3} + v_{to}}{609\cdot\sqrt{1 + \frac{15 + 1.98\cdot AA\cdot 10^{-3}}{273.15}}}\right)\cdot 5\,(1 + th_{aa})\,t_{aa} \tag{10.15}$$

For the ascent to cruise altitude phase, Time, t_{aa2c}, and distance, d_{aa2c}, are computed as in the deterministic case and Fuel consumption, fc_{aa2c}, is:

$$t_{aa2c} = \frac{CA_1 + \xi_{CA_1} - AA}{CR2C\cdot 60}$$

$$d_{aa2c} = v_{aa}\cdot 609\cdot\sqrt{1 + \frac{15 + 1.98\cdot(CA_1 + \xi_{CA_1})\cdot 10^{-3}}{273.15}}\cdot t_{aa2c} \cdot\cos\left(a\cdot\sin\left(\frac{\frac{CA_1 + \xi_{CA_1} - AA}{6076.115}}{v_{aa}\cdot 609\cdot\sqrt{1 + \frac{15 + 1.98\cdot(CA_1 + \xi_{CA_1})\cdot 10^{-3}}{273.15}}\cdot t_{aa2c}}\right)\right) \tag{10.16}$$

$$fc_{aa2c} = \left(2\xi_{ffr} - \left(CA_1 + \xi_{CA_1}\right)\cdot 10^{-3} + \xi_{w_{c1}}\cdot 10^{-3} + \left(v_{c1} + \xi_{v_{c1}}\right)\right)\cdot(1 + th_{aa})\,t_{aa2c}$$

ξ_{CA_1}: Uncertainty added to Cruise altitude 1 (ft).
$\xi_{w_{c1}}$: Uncertain wind at cruise level 1 (kt).
$\xi_{v_{c1}}$: Uncertainty added to speed at cruise level 1 (Mach).

for the first cruise phase, the distance, d_{c1}, is calculated as in the deterministic case. The time, t_{c1}, and Fuel consumption, fc_{c1}, of this first cruise phase are:

$$t_{c1} = \frac{d_{c1}}{\left(v_{c1} + \xi_{v_{c1}}\right)\cdot 609\cdot\sqrt{1 + \frac{15 + 1.98\cdot(CA_1 + \xi_{CA_1})\cdot 10^{-3}}{273.15}}}$$

$$fc_{c1} = \left(\left(2\xi_{ffr} - \left(CA_1 + \xi_{CA_1}\right)\cdot 10^{-3} + \xi_{w_{c1}}\cdot 10^{-3} + 5\left(v_{c1} + \xi_{v_{c1}}\right)\right)\cdot(1 + th_{aa}) + \left(100 - CI_{c1}\cdot 2\xi_{ffr}\right)\right)\cdot t_{c1} \tag{10.17}$$

For the intermediate phase, ascent/descent from cruise altitude 1 to cruise altitude 2, the time, t_{c2c}, the distance, d_{c2c}, and the fuel consumption, fc_{c2c}, are:

$$t_{c2c} = \frac{CA_1 + \xi_{CA_1} - CA_2}{CRCC\cdot 60}$$

$$d_{c2c} = \left(v_{c1} + \xi_{v_{c1}}\right)\cdot 609\cdot\sqrt{1 + \frac{15 + 1.98\cdot CA_2\cdot 10^{-3}}{273.15}}\cdot t_{c2c} \cdot\cos\left(a\cdot\sin\left(\frac{\frac{CA_1 + \xi_{CA_1} - CA_2}{6076.115}}{\left(v_{c1} + \xi_{v_{c1}}\right)\cdot 609\cdot\sqrt{1 + \frac{15 + 1.98\cdot CA_2\cdot 10^{-3}}{273.15}}\cdot t_{c2c}}\right)\right) \tag{10.18}$$

$$fc_{c2c} = \left(2\xi_{ffr} - CA_2 \cdot 10^{-3} + \xi_{w_{c2}} \cdot 10^{-3} + 5\left(v_{c1} + \xi_{v_{c1}}\right)\right) \cdot (1 + th_{aa})\, t_{c2c}$$

$\xi_{w_{c2}}$: Uncertain wind at cruise level 2 (kt)

The second cruise section of the flight is defined by the distance, d_{c2}, which is computed as in the deterministic case, the time, t_{c2}, and the fuel consumption, fc_{c2}, computed as:

$$t_{c2} = \frac{d_{c2}}{(v_{c1}+\xi_{v_{c1}}) \cdot 609 \cdot \sqrt{1+\frac{15+1.98 \cdot CA_2 \cdot 10^{-3}}{273.15}}}$$

$$fc_{c2} = \left(\left(2\xi_{ffr} - CA_2 \cdot 10^{-3} + \xi_{w_{c2}} \cdot 10^{-3} + 5\left(v_{c1} + \xi_{v_{c1}}\right)\right) \cdot (1 + th_{aa}) + \left(100 - CI_{c2} \cdot 2\xi_{ffr}\right)\right) \cdot t_{c2} \tag{10.19}$$

Finally, the descent to final approach altitude is defined by the time, t_{d2a}, calculated as in the deterministic case, the distance, d_{d2a}, and the tuel consumption, fc_{d2a}. They are computed as:

$$d_{d2a} = \left(v_{c1} + \xi_{v_{c1}}\right) \cdot 609 \cdot \sqrt{1 + \frac{15+1.98 \cdot (h_d+3000) \cdot 10^{-3}}{273.15}} \cdot t_{d2c} \cdot \cos\left(a \cdot \sin\left(\frac{\frac{CA_2-(h_d+3000)}{6076.115}}{(v_{c1}+\xi_{v_{c1}}) \cdot 609 \cdot \sqrt{1+\frac{15+1.98 \cdot (h_d+3000) \cdot 10^{-3}}{273.15}} \cdot t_{d2c}}\right)\right) \tag{10.20}$$

$$fc_{d2a} = \left(2\xi_{ffr} - CA_2 \cdot 10^{-3} + \xi_{w_{c2}} \cdot 10^{-3} + 5\left(v_{c1} + \xi_{v_{c1}}\right)\right)(1 + th_{aa})\, t_{d2a}$$

The calculation of the time and the distance for each flight phase is performed as in the deterministic case. Flight parameters are applied. Simplifications, as defined in the deterministic case, have also been applied in the stochastic and robust design. Optimization search space is the same range of values already described in the deterministic case section.

The definition of the uncertainties is listed below. It does not matter if the uncertain parameter is defined as a perturbation of the parameters (a addition to the parameters, from the mathematical point of view), or directly as the value to be used in the equation.

ξ_{ffr}: uncertain fuel flow (kg/min) Gaussian distribution, Mean = 1000, Standard deviation = 0.5

$\xi_{w_{to}}$: uncertain wind during take-off (kt) Gaussian distribution, Mean = 15, Standard deviation = 0.1

$\xi_{w_{c1}}$: Uncertain wind at cruise level 1 (kt) Uniform range = [−10; 60]

$\xi_{w_{c2}}$: Uncertain wind at cruise level 2 (kt) Uniform range = [−15; 40]

ξ_{CA_1}: Uncertainty added to Cruise altitude 1 (ft) Uniform range = [1500; 2500]
$\xi_{v_{c1}}$: Uncertainty added to speed at cruise level 1 (Mach) Gaussian distribution, Mean = −0.02, Standard Deviation = 0.02

10.5.1.1 Stochastic Definition

The minimization of the equations (10.21) defines the stochastic optimization problem. It considers the mean values of the quantities of interest.

$$FC_s = \mu(fc_{tx} + fc_{to} + fc_{a2aa} + fc_{aa} + fc_{aa2c} + fc_{c1} + fc_{c2c} + fc_{c2} + fc_{d2a} + fc_{fa})$$

$$T_s = \mu(t_{tx} + t_{to} + t_{a2aa} + t_{aa} + t_{aa2c} + t_{c1} + t_{c2c} + t_{c2} + t_{d2a} + t_{fa}) \tag{10.21}$$

$$D_s = \mu(d_{tx} + d_{to} + d_{a2aa} + d_{aa} + d_{aa2c} + d_{c1} + d_{c2c} + d_{c2} + d_{d2a} + d_{fa})$$

10.5.1.2 Robust Design Definition

The robust design problem has been defined as the minimization of the following Eq. (10.22), which considers both the mean and the standard deviation of the quantities of interest.

$$FC_r = \mu(fc_{tx} + fc_{to} + fc_{a2aa} + fc_{aa} + fc_{aa2c} + fc_{c1} + fc_{c2c} + fc_{c2} + fc_{d2a} + fc_{fa})$$

$$T_r = \mu(t_{tx} + t_{to} + t_{a2aa} + t_{aa} + t_{aa2c} + t_{c1} + t_{c2c} + t_{c2} + t_{d2a} + t_{fa})$$

$$D_r = \mu(d_{tx} + d_{to} + d_{a2aa} + d_{aa} + d_{aa2c} + d_{c1} + d_{c2c} + d_{c2} + d_{d2a} + d_{fa}) \tag{10.22}$$

$$FC_{2r} = \sigma(fc_{tx} + fc_{to} + fc_{a2aa} + fc_{aa} + fc_{aa2c} + fc_{c1} + fc_{c2c} + fc_{c2} + fc_{d2a} + fc_{fa})$$

$$T_{2r} = \sigma(t_{tx} + t_{to} + t_{a2aa} + t_{aa} + t_{aa2c} + t_{c1} + t_{c2c} + t_{c2} + t_{d2a} + t_{fa})$$

10.6 Results

10.6.1 Results of the Deterministic Test Case

The deterministic test case provides results showing the best route, see Fig. 10.6. This route is the best one without any other consideration. This situation occurs because the way-points defined clearly identify it. The mission profile is not producing any effect, so the route is not modified.

The Pareto front, more precisely its 3D representation, is shown in Fig. 10.3 including the three objective functions. The Pareto front is in reality a 2D curve because distance could be discarded as it keeps constant to the best optimal route distance (Fig. 10.4).

From the Pareto front, one can extract the mission profile to get a more application-specific results. The mission profiles of the optimal individuals can be found on Fig. 10.5. The climbing stage to the cruise level is related to the cruise altitude, so in this way the fuel consumption is minimized. Each optimal individual is showing a different combination of the objective functions, producing a different combination on the design parameters. A relevant example is the one with the steepest climb. Due to the fact a steep climb means higher fuel consumption, it is compensated with the a very low cruise level. The optimizer is finding the best trade off according the objective functions values and the restrictions applied.

Some of the solutions, which seems not to be very realistic due to a poor consideration of air traffic control restrictions, show long descent phases before final approach. The simplifications and approximations accepted at the very beginning have lead to these kind of solutions (Fig. 10.5).

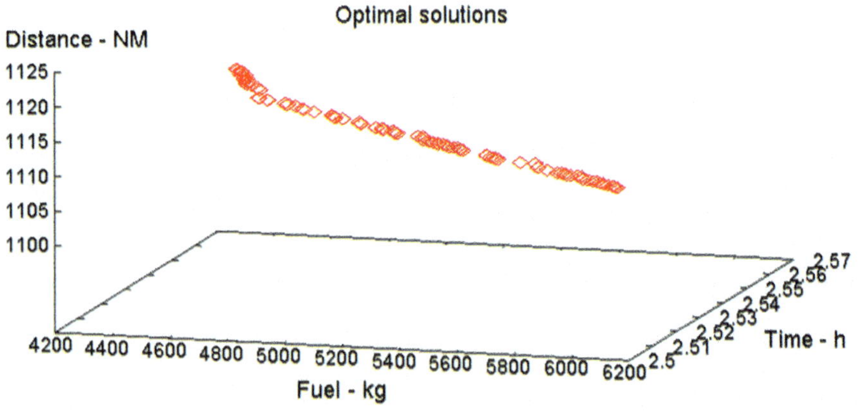

Fig. 10.3 Set of optimal solutions for the three objective functions of the deterministic case, 3D view

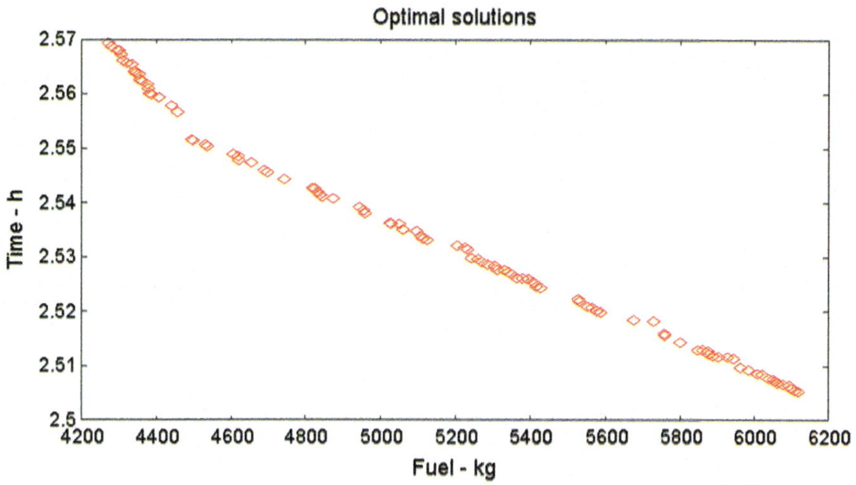

Fig. 10.4 Set of optimal solutions for the three objective functions of the deterministic case

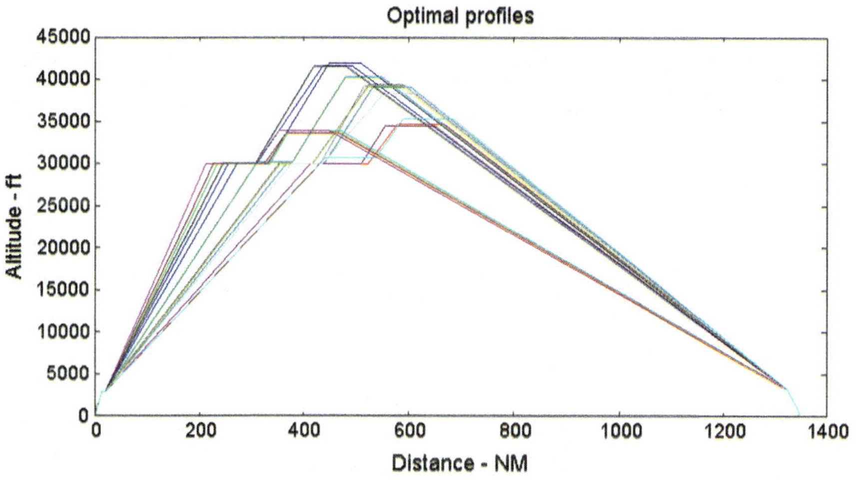

Fig. 10.5 Mission profiles for the best route

10.6.2 Results of the Stochastic Test Case

An optimal solution have been obtained, which defines a unique optimal distance. It is the same situation as in the previous analysis with the deterministic definition. In both stochastic and robust optimization cases the distance is slightly modified with the inclusion of uncertain parameters. Then, as it happens before, the few options for each way-point to calculate the route produces this undesired effect. Further improvements should be applied in the definition of the route to ensure that several

Fig. 10.6 Best route

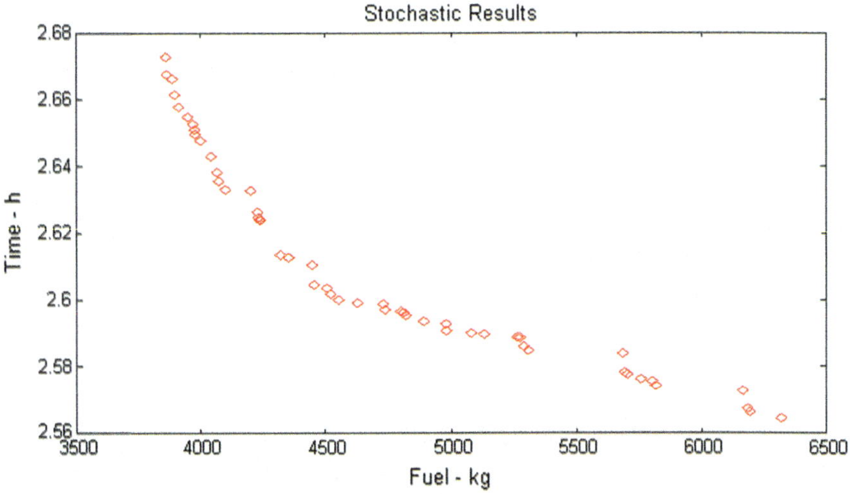

Fig. 10.7 Pareto Front for the stochastic analysis

optimal routes are obtained according the uncertain conditions. As the distance can be discarded as result of interest in the Pareto front, Fig. 10.7 shows the best trade-off solutions of the Fuel and Time objective functions obtained with the stochastic procedure. To get a clear representation of the optimal solutions, Fig. 10.8 describes the mission profiles of the optimal individuals. The definition of uncertainty leads to a combination of steeper climb ratio with higher cruise altitude. The stochastic case is not considering the robustness of the solutions, because it is not considering the standard deviation as an objective function or part of it. The robust design procedure solves this issue.

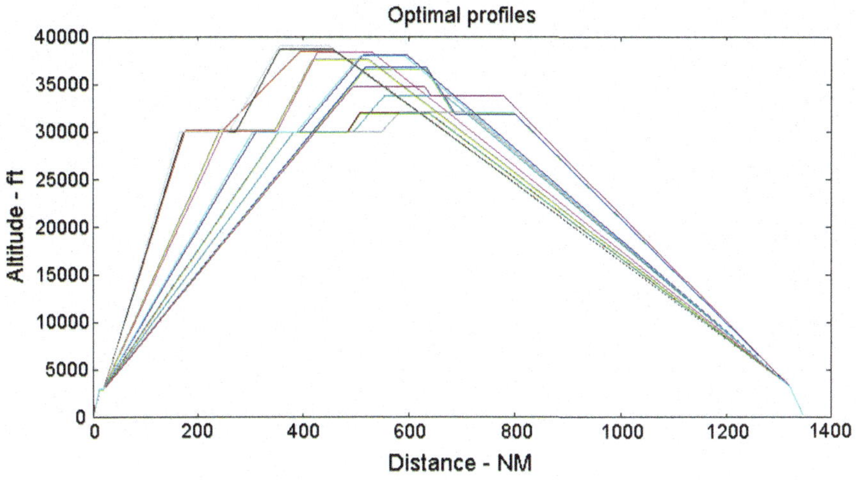

Fig. 10.8 Optimal Profiles for the stochastic analysis

10.6.3 Results of the Robust Test Case

Robust design considers 5 objective functions, including two variance measures (the standard deviation). It means that it will be easy to understand the variability associated to the uncertain design parameters, and their effect on the objective functions. The Pareto front of the solution is shown in Fig. 10.9. The best trade-offs between fuel and time objective functions are plotted. The shape of the Pareto front cannot be

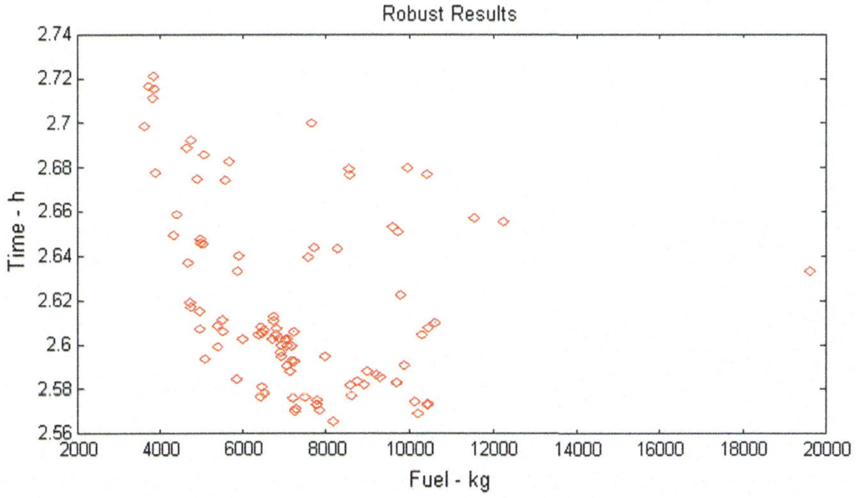

Fig. 10.9 Pareto Front for robust optimization analysis

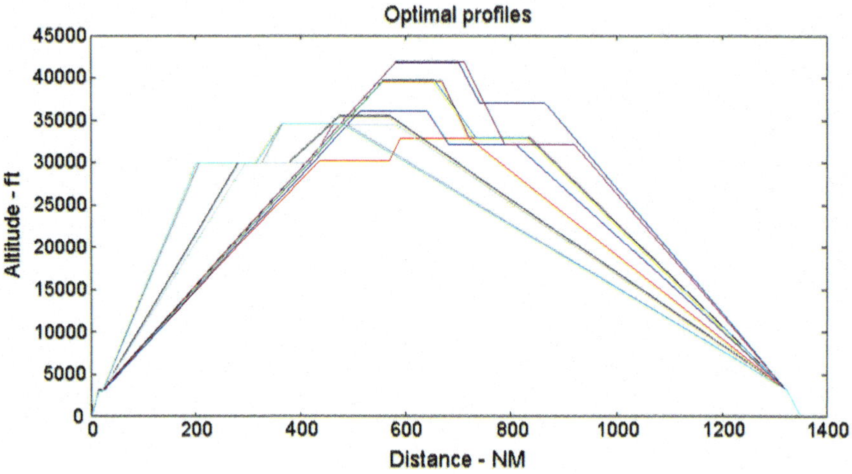

Fig. 10.10 Optimal Profiles for the robust optimization analysis

represented by a 2D plot due to the definition of 5 objective functions. Even 3D representation are not good capturing the real shape of the Pareto Front, the dimension of the front if a 5D object. For comparison purposes, the time and the fuel consumption have been plotted in the pseudo-Pareto front. Then an approximate comparison with the previous results can be performed. The robust definition, using the standard deviation, means that the optimal individuals are hardly comparable to the previous ones. Values of the functions are similar, but the distribution of them along the Pareto front is different. Small differences can be detected in Fig. 10.10 in comparison with Fig. 10.8, while comparing with Fig. 10.9 the values of climb rate and descent rates are not spread so regularly than in previous cases.

10.7 Conclusions

Two major issues must be highlighted; the first one is the simplifications applied to the definition of the problem. The second one is the partial information used in the definition of the models, mainly fuel consumption model, due to the restrictions applied by the air carrier. Anyway, the results are accurate enough. The general behavior of the problem and the solution is accepted by experts in the air carrier company as accurate enough.

Regarding the results, the presence of a unique optimal distance, has simplified the analysis of a multi dimensional solution space, from a 3D optimal space, it has been reduced to a 2D space. It only considers the time and the fuel consumption as objective functions of interest. An improvement to avoid this situation is to define a larger number of option to select in each way-point of the route, plus defining closer

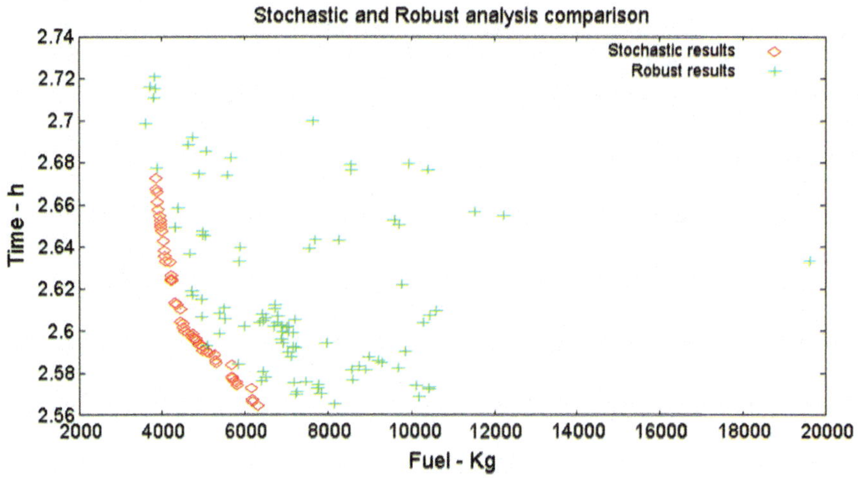

Fig. 10.11 Two non-deterministic case comparison

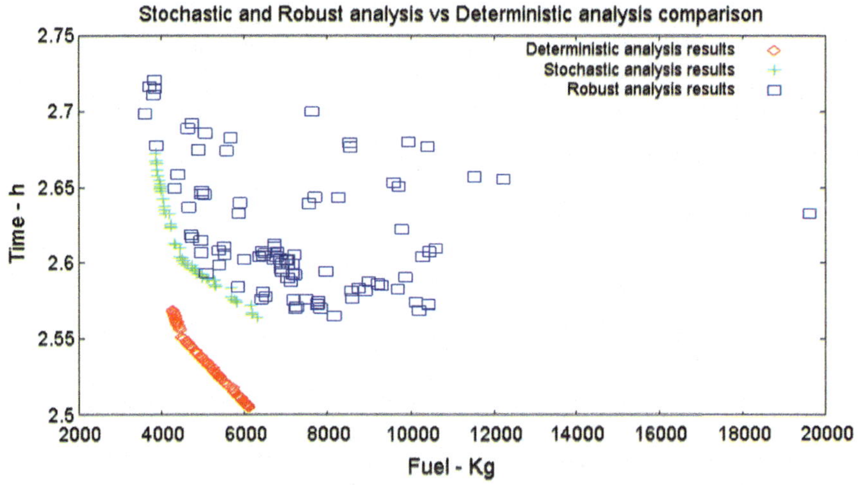

Fig. 10.12 Deterministic and non-deterministic cases comparison

routes. Although the clear limitations of the analysis, it has been an opportunity to validate and compare the definition and application of the three methodologies in a problem with real industrial interest. This problem has a close relationship with the objectives of the 2020 Vision strategy described in (Busquin et al. 2001). This strategy, confirmed and expanded in the FlightPath2050 document (Darecki et al. 2011), defines the aim of a very important emission reduction (directly related to fuel consumption), as well as an important noise reduction.

A comparison of the results from the three methodologies is shown in Figs. 10.11 and 10.12. The first one compares the results obtained in the stochastic and the robust optimization procedures. The two Pareto Front show some similarities, considering that the robust one is a projection to the Time-Fuel plane. The stochastic front is limiting the robust one, which demonstrates the agreement between to the set of results. As mentioned, the robust definition, which defines 5 objective functions, and the standard deviation as two of them, introduces new solutions in a multi-dimensional solution space.

Adding the deterministic results to the comparison, as as shown in Fig. 10.12, the reader can realize how working with the value of interest or its mean modify the solution. One can thing that the deterministic solution is obtaining a better solution, minimizing the fuel and the time. Nothing can be claimed in regards of the robustness of the solution. The comparison between the methodologies should take into account this important issue about the uncertainty definition and effect to the results. The simplified model, which is based on mathematical relationship of the variables, facilitated the computation of the individuals. Its associated computational cost is low. Opposite to the other applications like CFD or structural mechanics analysis, the actual Mission planning application does not present such a huge difference between the computational cost of the deterministic and the robust definition. Although the robust case is more expensive, now, the cost is not a limiting issue. The improvement of the model, while deleting the simplifications and approximations applied here, can lead to a very complex definition, which will include differential equations. Its solution can be costly compared to the problem described here, so the difference between the three cases can be larger in computational time.

References

AIRBUS (2008) Getting the grip with; airbus A320 family, performance retention and fuel savings. Flight operations support and services, Airbus, Toulouse, France

BOEING (2007) Fuel conservation strategies: cost index explained. Boeing aero quarterly, Seattle, USA

Busquin P et al (2001) European aeronautics: a vision for 2020; meeting society's needs and winning global leadership. Report of the group of personalities, European Commission, Luxembourg

Darecki M, Edelstenne C, Enders T, Fernandez E, Hartman P, Herteman J-P, Kerkloh M, King I, Ky P, Mathieu M, Orsi G, Schotman G, Smith C, Worner J-D (2011) Flightpath 2050: Europe's vision for aviation maintaining global leadership and serving society's needs. Technical Report of the high level group on aviation research, European Commission, Brussels

Deb K (2001) Multi-objective optimization using evolutionary algorithms, vol 16. Wiley, New Jersey

Deb K, Agrawal S, Pratap A, Meyarivan T (2000) A fast elitist non-dominated sorting genetic algorithm for multi-objective optimization: NSGA-II. In: International conference on parallel problem solving from nature. Springer, Heidelberg, pp 849–858

Deb K, Pratap A, Meyarivan T (2001) Constrained test problems for multi-objective evolutionary optimization. In: International conference on evolutionary multi-criterion optimization. Springer, Heidelberg, pp 284–298

Deb K, Anand A, Joshi D (2002a) A computationally efficient evolutionary algorithm for real-parameter optimization. Evol Comput 10(4):371–395

Deb K, Pratap A, Agarwal S, Meyarivan T (2002b) A fast and elitist multiobjective genetic algorithm: NSGA-II. IEEE Trans Evol Comput 6(2):182–197

Gonzalez LF, Lee DS, Walker RA, Periaux J (2011) Aircraft emission reduction through multidisciplinary flight path optimisation. In: Love D (ed) The 14th Australian international aerospace congress (AIAC14). Melbourne Convention Centre, Melbourne, VIC: Waldronsmith Management - AIAC

Lee DS, Periaux J, Gonzalez LF, Srinivas K, Onate E (2012) Robust multidisciplinary UAS design optimisation. Struct Multidiscip Optim 45(3):433–450

Part III
Translational Research

Chapter 11
Reallocation of Logistics Costs in a Cooperative Network of Sawmills

Patrik Flisberg, Mikael Frisk, Mario Guajardo and Mikael Rönnqvist

Abstract While collaborative logistics has the potential to provide savings to organizations, the individual result of sub-units within an organization might not directly benefit from the collaboration. This cause problems as the sub-units may be their own result or profit centers. We face this problem in the context of an organized network of sawmills. The organization benefits from timbering exchange and joint transports with an external company. The collaboration with this external company, however, implies an increase in the direct cost of supplying some of the sawmills. This occurs because some of the flows which would be used to supply these sawmills in absence of the collaboration, are assigned to the external company in the collaborative solution. In order to make the collaboration profitable for all sawmills, the organization must reallocate the cost among the different members in the network. We address this problem by using concepts of cooperative game theory. We apply these concepts in a case involving a network of 12 Scandinavian sawmills which together cooperate with an external procurement company. The collaboration in this case results in 3.3% savings for the network. A slightly modified version of the equal profit method allows to reallocate the cost of the collaborative solution in such a way that the cost of all sawmills is reduced with respect to their direct cost in absence of collaboration, while also assures the stability of the cooperation.

P. Flisberg · M. Frisk
The Forestry Research Institute of Sweden, Uppsala, Sweden
e-mail: pafli@mweb.co.za

M. Frisk
e-mail: konsult@mikaelfrisk.se

M. Guajardo (✉)
Department of Business and Management Science,
NHH Norwegian School of Economics, Helleveien 30, 5045 Bergen, Norway
e-mail: mario.guajardo@nhh.no

M. Ronnqvist
Département de génie mécanique, Université Laval, Québec, QC G1V 0A6, Canada
e-mail: mikael.ronnqvist@gmc.ulaval.ca

P. Diez et al. (eds.), *Computational Methods and Models for Transport*,
Computational Methods in Applied Sciences 45,
DOI 10.1007/978-3-319-54490-8_11

11.1 Introduction

Collaborative logistics has received increasing attention in theory and practice, due to its potential to benefit organizations and to develop sustainability. Primary logistics activities, such as transportation and inventory management, can be improved by combining the needs of different companies. The forestry industry has served as a venue for several works in this stream. Potential cost savings from about 6 to 22% have been estimated in forestry related cases of collaborative logistics studied by Audy et al. (2011), Flisberg et al. (2015), Frisk et al. (2010), Guajardo et al. (2016) and Guajardo and Rönnqvist (2015). In addition, negative environmental impact of emissions can be reduced by comparable numbers. Two main problems have been studied in such references. One deals with how to structure the coalitions and the other one with how to allocate the costs or savings to the organizations that participate in the collaboration. While these are important problems foreseing the implementation of collaboration, an alternative problem, more absent in the literature, is to study the effect of collaboration in the outcome of different sub-units of the partners. This is the view adopted in Flisberg et al. (2016), motivated by a real case in forestry arising in Sweden. In such case, collaboration between an organized network of sawmills and an external company takes place by means of wood bartering and joint transportation plans. The overall collaborative solution is better for the organization of sawmills, in the sense that the total logistics cost is reduced in comparison to the non-collaborative solution. However, the outcome of some of the sawmills in the network is worsened, because the wood they would use in the non-collaborative solution is assigned to the external company in the collaborative solution. The reason is that the sawmills pay for the delivered wood directly to hauliers. Flisberg et al. (2016) studied several scenarios pointing out large differences can occur in the outcomes of the sawmills with or without presence of an external collaborating company. This motivates the need for reallocating the costs among the sawmills in such a way that all of them benefit from the collaboration. In this article, we intend to address this need by using cost allocation methods taken from the cooperative game theory literature. The use of this methods in collaborative transportation has received increasing attention in recent years, as reviewed by Guajardo and Rönnqvist (2016).

11.2 Background

In this section, we introduce some preliminary settings and examples which are useful to illustrate the problem faced by the collaborative network of sawmills.

11.2.1 Transportation Planning

Consider the problem of transporting a single product, where a company has a set of supply points I and must satisfy demand of a set of points J. The maximum supply quantity at point $i \in I$ is s_i and the demand quantity at point $j \in J$ is d_j. The unit cost of flow between supply point i and demand point j is c_{ij}. The objective of the company is meeting demand at minimum cost. A straightforward solution to this problem can be found by solving the following linear programming model in the flow variables y_{ij}.

$$\min\ C = \sum_{i \in I} \sum_{j \in J} c_{ij} y_{ij} \tag{11.1}$$

$$\text{s.t.} \sum_{j \in J} y_{ij} \leq s_i \quad \forall i \in I \tag{11.2}$$

$$\sum_{i \in I} y_{ij} = d_j \quad \forall j \in J \tag{11.3}$$

$$y_{ij} \geq 0 \quad \forall i \in I, j \in J \tag{11.4}$$

The above model addresses a classic transportation problem studied since early times (Hitchcock 1941; Tolstoi 1930). Standard extensions are to include several products and time periods which also require inventory considerations.

Given the value of the flow variables y_{ij}, we refer by K_j to the *direct cost* of fulfilling the demand point j, that is,

$$K_j = \sum_{i \in I} c_{ij} y_{ij}. \tag{11.5}$$

In the non-collaborative approach a single company solves its own transportation problem. In collaboration, the supply points of a company can be used to fulfil demand from another company, which is handled by defining sets I and J as the union of the supply points and demand points of all the companies, respectively. The optimal solution for a coalition of companies usually provides some savings with respect to the sum of the cost of the individual plans, as the supply points of one company may be closer to the demand points of another one and vice-versa. This, however, does not imply that the direct cost of fulfilling demand decreases for all points within a network. We illustrate this situation by two examples below.

Example 11.1 Consider the example in Fig. 11.1. Company A has two demand nodes (d_{A_1} and d_{A_2}) and one supply node (s_{A_1}). Company B has one demand node (d_{B_1}) and one supply node (s_{B_1}). The unitary transportation cost on each arc is denoted by c. Suppose the demand quantities are $d_{A_1} = 5$, $d_{A_2} = 5$, $d_{B_1} = 10$, and the supply quantities $s_{A_1} = 10$, $s_{B_1} = 10$. The dashed lines and continuous lines in Fig. 11.1 represent the demand fulfilments in the non-collaborative and collaborative case, respectively. For a given solution, we refer by C_n to the cost of company n (that is, the sum of the direct costs of fulfilling the demand of all points of company n), and by C_{total} to the sum of the costs of all companies. We add a bar in this notation for the collaborative solution, to distinguish it from the non-collaborative solution.

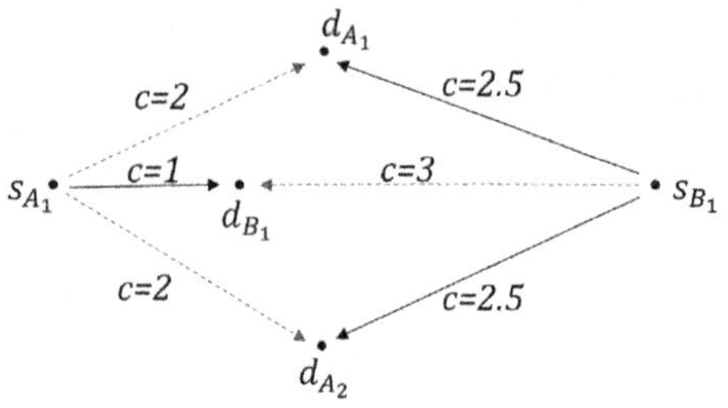

Fig. 11.1 Network of supply and demand points, and transportation costs for Example 11.1

Non-collaborative solution. The direct costs at demand points, the total direct cost of the companies and the total costs when companies A and B do not collaborate are calculated as follows:

$K_{A_1} = 2 \times 5 = 10$
$K_{A_2} = 2 \times 5 = 10$
$K_{B_1} = 3 \times 10 = 30$
$C_A = K_{A_1} + K_{A_2} = 20$
$C_B = K_{B_1} = 30$
$C_{total} = C_A + C_B = 50$

Collaborative solution. The direct costs at demand points, the total direct cost of the companies and the total costs when companies A and B collaborate are calculated as follows:

$\bar{K}_{A_1} = 2.5 \times 5 = 12.5$
$\bar{K}_{A_2} = 2.5 \times 5 = 12.5$
$\bar{K}_{B_1} = 1 \times 10 = 10$
$\bar{C}_A = \bar{K}_{A_1} + \bar{K}_{A_2} = 25$
$\bar{C}_B = \bar{K}_{B_1} = 10$
$\bar{C}_{total} = \bar{C}_A + \bar{C}_B = 35$

In this example, the collaborative situation reduces the total cost by 15, but the total direct cost perceived by company A increases by 5.

Example 11.2 Consider a second example illustrated in Fig. 11.2. This situation is similar to the one described in Example 11.1, but the costs in some arcs are slightly modified.

Non-collaborative solution. The direct costs at demand points, the total direct cost of the companies and the total costs when companies A and B do not collaborate are calculated as follows:

$K_{A_1} = 1.5 \times 5 = 7.5$
$K_{A_2} = 2.6 \times 5 = 13$
$C_A = K_{A_1} + K_{A_2} = 20.5$
$C_B = 3 \times 10 = 30$
$C_{total} = C_A + C_B = 50.5$

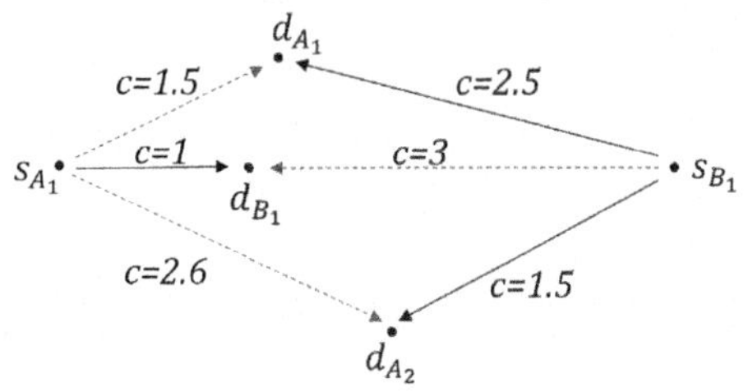

Fig. 11.2 Network of supply and demand points, and transportation costs for Example 11.2

Collaborative solution. The direct costs at demand points, the total direct cost of the companies and the total costs when companies A and B collaborate are calculated as follows:

$\bar{K}_{A_1} = 2.5 \times 5 = 12.5$
$\bar{K}_{A_2} = 1.5 \times 5 = 7.5$
$\bar{C}_A = \bar{K}_{A_1} + \bar{K}_{A_2} = 20$
$\bar{C}_B = 1 \times 10 = 10$
$\bar{C}_{total} = \bar{C}_A + \bar{C}_B = 30$

In this example, the collaborative situation reduces the total cost by 20.5 and the direct cost perceived by company A decreases by 0.5. The cost of the demand point A_1, however, increases by 5.

11.2.2 Cooperative Network of Sawmills

In the previous two examples, the collaboration achieves savings with respect to the non-collaborative solution, but leaves some demand centres or a full company with higher direct cost. This is the situation faced by an organized network of sawmills which primarily motivates our work. The sawmills belong to a number of owners. The owners have organized a common purchasing and transportation organization to improve their logistics operations. Each sawmill can be seen as an independent profit center within the whole organization. The sawmills demand wood by different assortments, which is used as raw material in their production process. The wood is procured from a number of supply areas. Several transport operations are involved in carrying the wood from the supply areas to the sawmills. When transports are done each sawmill is invoiced the actual cost of the truck transportation. Excess costs for train transport, loading at the terminal and transport to the terminal are charged evenly to at all volumes. The reason for this handling is to give some freedom in controlling the train flows.

The purchasing organisation re-distributes the actual costs of purchasing wood. If one sawmill must purchase an expensive volume (a forest harvest area) and another a less expensive one, there is a model to compute the average cost. This average cost is then used to re-distribute the purchasing cost so that all sawmills pay the same average price. There is an objective to process the logistic cost in a similar process.

The organization can collaborate with external companies by wood bartering. This is usually good for the organization seen as a whole, as such trading agreements often involve large savings. However, individual sawmills may be both victims and beneficiaries of wood bartering. Sawmills near the volumes made available from an external company tend to have shorter transport distances, which is what demand point A_2 experiences in Fig. 11.2. On the other hand, sawmills near the removed bartered volumes tend to have longer transport distances, which is what demand point A_1 experiences in Fig. 11.2.

The control of transports is based on a supply plan to the sawmills where the volumes from different suppliers are balanced against the demand quantities. Supply plans are worked out in the budget process through a dialogue with all suppliers. The supply plan is then broken down into monthly plans. The starting point for this planning is to minimize the total transportation cost.

Since each sawmill is an independent profit center, some questions naturally arise regarding the logistics costs. One of these is how to fairly re-distribute the logistics costs among the network. As more firms or sub-units work together, it is important that common objectives and requirements are clear. Cooperation in logistics requires a large transparency of information and description of how the distribution of cost reductions (or gains) occurs. Critical for a distribution is that it is seen as fair by all entities or owners. Henceforth, our aim is to study how solution concepts from cooperative game theory can deal with the problem of reallocating costs among the different sawmills.

11.3 Cooperative Game Theory Concepts

In this section, we summarize some cooperative game theory concepts. These refer to some classic literature on transferrable utility games, which have gained increasing attention from the literature on logistics more recently.

11.3.1 Basic Concepts

Let $N = \{1, \ldots, n\}$ be the set of all players or so called *grand coalition*. Let T be the set of all non-empty subsets of N. The characteristic function $C : T \to \mathbb{R}$ assigns to each coalition S in T the *cost* of coalition S. The pair (N, C) defines a transferable utility game. A *preimputation* or *cost allocation vector* $x = (x_1, \ldots, x_n)$ assigns to each player j in N a quantity

$x_j \in \mathbb{R}$ such that

$$\sum_{j \in N} x_j = C(N). \tag{11.6}$$

Equation (11.6), known as efficiency condition, states that the cost of the grand coalition N is split among its members according to the allocation x. An *imputation* is a preimputation which satisfies the following condition:

$$x_j \leq C(\{j\}) \forall j \in N \tag{11.7}$$

Inequalities (11.7) are known as the *individual rationality* conditions, which state that the cost allocated to each player is not greater than the cost it would perceive by standing alone. The same principle can be stated in general for all coalitions which can be formed among the set of players. This is captured in the *rationality* conditions, which are formulated as follows:

$$\sum_{j \in S} x_j \leq C(S) \forall S \in T \tag{11.8}$$

The *core* of the game is the set of preimputations that satisfy rationality. A cost allocation vector that belongs to the core is said to be *stable*. Stability in this context means that no subset of players have incentives to break away from the grand coalition.

11.3.2 Cost Allocation Methods

A cost allocation method is a function or procedure to compute a cost allocation vector for a given game (N, C). Many methods have been proposed in the literature. We limit ourselves to study three of them and refer the reader to Guajardo and Rönnqvist (2016) for an overview on a variety of other methods which have been used in collaborative transportation.

Shapley values

The allocation by Shapley (1953) is one of the most used concepts in cooperative games. This concept allocates to each player j an average of the marginal costs it implies when entering coalitions, as follows:

$$x_j = \sum_{S \subseteq N : j \in S} \frac{(n - |S|)!(|S| - 1)!}{n!} \cdot [C(S) - C(S \setminus \{j\})] \quad \forall j \in N \tag{11.9}$$

The Shapley allocation is the only one satisfying four celebre axioms: efficiency, symmetry, dummy property and additivity. Another appealing feature of this allocation is that it is unique.

Nucleolus

Introduced in a seminal work by Schmeidler (1969), the nucleolus is to date one of the most used allocation methods in cooperative game theory. Originally defined in the form of a profit game, we present its definition adapted to a cost sharing game (N, C). Define the excess of coalition S at x in game (N, C) as $\varepsilon(x, C, S) = C(S) - \sum_{j \in S} x_j$. The excess of a coalition S at an allocation x can be interpreted as a measure of satisfaction of the coalition with this allocation. The larger the excess of S, the more satisfied coalition S is, in the sense that it achieves larger savings. For a game (N, C), define the excess vector at x as $e(x, C) = (\varepsilon(x, C, S_1), \ldots, \varepsilon(x, C, S_p))$, where the sets S_i represent the coalitions in $K \setminus \{N\}$ and $p = 2^n - 2$. Denote by $\mathcal{X}$ to the set of imputations. The nucleolus $\mathcal{N}$ of the cost sharing game (N, C) can be defined as $\mathcal{N} = \{x \in \mathcal{X} : \theta(e(x, C)) \succeq \theta(e(y, C)) \forall y \in \mathcal{X}\}$. Thus, the nucleolus is the set of imputations that lexicographically maximizes the excess vector. Appealing well-known features of the nucleolus is that it is unique and that it always belong to the core whenever this is non-empty. Its computation is a bit more complex than applying a single formula as the Shapley values. Several algorithms have been proposed in the literature. We refer the reader to Fromen (1997) for an algorithm and to Guajardo and Jörnsten (2015) for several numerical examples on how to use it correctly.

Equal profit method

This method was introduced by Frisk et al. (2010), in light of the difficulties of making companies to agree on allocations providing them with different relativesavings with respect to each other. If company j is allocated a cost x_j, the savings relative to its stand alone cost are $\frac{C(\{j\}) - x_j}{C(\{j\})}$. Thus, the difference between the relative savings of companies j and i is $\frac{x_i}{C(\{i\})} - \frac{x_j}{C(\{j\})}$. The equal profit method (EPM) attempts to find an allocation in the core such that the maximum of the differences between relative savings is minimized. We propose here a slightly modified version, referred as EPMm, in which the relative savings are not computed with respect to the stand alone costs but rather with respect to the direct cost K_j perceived by the sawmills in absence of collaboration. This is because in the case that motivates our research, the sawmills already collaborate within their organization. The conflict arises when the organization collaborates with the external company. Thus, instead of using the stand alone cost as basis to compute the savings, what turn more important for comparison among the sawmills are their savings with respect to the direct costs they perceive without collaborating with the external company. With this distinction, the modified equal profit method allocation can be found by solving the following linear programming model:

$$\min f \tag{11.10}$$

$$\text{s.t.} \quad f \geq \frac{x_i}{K_i} - \frac{x_j}{K_j} \ \forall\, i, j \in N \tag{11.11}$$

$$\sum_{j \in S} x_j \leq C(S) \ \forall\, S \subset N \tag{11.12}$$

$$\sum_{j \in N} x_j = C(N) \tag{11.13}$$

$$f \in \mathbb{R} \ , \ x_j \in \mathbb{R} \ \ \forall\, j \in N \tag{11.14}$$

Objective function (11.10) minimizes f and constraints (11.11) impose f as an upper bound on all pairwise differences between savings of the sawmills with respect to their direct cost without collaboration with the external company. Thus, (11.10) and (11.11) together provide that f equals the maximum of these differences. Constraints (11.12) and (11.13) impose rationality and efficiency, respectively. Then, if a solution to the model is obtained, it is an allocation in the core. If the core is empty, this model is infeasible. Constraints (11.14) state the decision variables are continuous, which we have included explicitly to emphasize that some players could get a negative allocation, that is, a payment instead of a cost.

11.4 Case Study

Our case study is based on a main Swedish organization accounting for a network of 12 sawmills. We will refer by N to the organization and by $N_1, \ldots, N_{12}$ to each of the sawmills in the network. The organization has full supply responsibility for nine sawmills in Sweden and partially for three sawmills in Norway. The purchasing areas for raw material comprises central Sweden from the Norwegian border in the west to the eastern coast. The industries are relatively scattered in the geography. The procurement involves approximately 4000 harvest areas annually. N collaborates with an external company which we will refer by E.

11.4.1 Data Features and Direct Costs

The data comprise all logistics operations on a particular month. This includes journeys made by road and rail with information about the quality, volume, location of assets and costs. This applies to the organization of sawmills as well as for another main company involved in collaboration. The collaboration consists of wood bartering by different assortments and the deployment of joint transportation plans. The corresponding flows involved in the operations by N amounted to 281,624 m^3. About 14,000 m^3 of this volume was involved in the wood bartering with E. We use the data as input in a decision support system called FlowOpt. The system minimizes the total cost of timber flows, based on an optimization model. The model incorporates some more sophisticated features into the classic transportation planning problem formulated in (11.1)–(11.4). For example, it allows for different means of transportation, such as truck, train and vessel. It can also be used to coordinate the exchange of wood between several companies. The system utilizes the extended forest version of the Swedish National Road Database (NVDB) called SNVDB to calculate the correct distance between the regions and the receiver terminals.It also uses data on the transport cost functions for the input ranges, the network of terminals, train systems and costs. A more detailed description of DSS FlowOpt and the underlying optimization model can be found in Flisberg et al. (2016) and Forsberg et al. (2005).

Table 11.1 Direct cost allocation (in kSEK) to the sawmills with and without collaboration with the external company E

Sawmill	K_j (direct cost without E)	$\bar{K}_j$ (direct cost with E)	Variation (%)
N_1	4,057	3,770	−7.1
N_2	2,377	2,173	−8.6
N_3	2,073	2,013	−2.9
N_4	43	43	0.0
N_5	1,251	1,222	−2.3
N_6	699	699	0.0
N_7	1,670	1,604	−3.9
N_8	678	722	6.4
N_9	300	349	16.2
N_{10}	101	101	0.0
N_{11}	2,004	2,010	0.3
N_{12}	1,715	1,699	−0.9
Total	**16,968**	**16,405**	**−3.3**

The direct costs of each sawmill with and without collaboration with the external company E are shown in Table 11.1. In the collaborative solution, the cost perceived by the organization of sawmills is in total 16,405 kSEK. That is 563 kSEK or 3.3% less than the cost it would perceive without collaborating with the external company E. The effect of the collaboration in the individual result of the sawmills differ. For six of them, the direct cost is reduced. The reduction ranges from 0.9%, in the case of N_{12}, to 8.6% for N_2. Three sawmills (N_4, N_6 and N_{10}) are not affected by the collaboration. The remaining three sawmills, on the other hand, perceives an increase in the direct costs due to the collaboration. While this increase is a moderate 0.3% for N_{11}, it is a more pronounced 6.4% for N_8 and a considerably high 16.2% for N_9.

11.4.2 Cost Allocations

In order to apply the cost allocation methods introduced in Sect. 11.3, we need to define the characteristic function. We define its values as $C(N)$ equal to the total cost perceived by N in collaboration with E (that is, $C(N) = 16,405$) and $C(S)$ the cost of optimizing the plan for the samwills in $S \subset N$ without collaboration with E. Note the characteristic function must be computed for each of the non-empty coalitions which can be formed among the 12 sawmills, that is, $2^{12} - 1 = 4095$. For each of these coalitions, we run the corresponding instance in the DSS FlowOpt, which provides us with the value of $C(S)$. Then we apply the cost allocation methods and obtain the results shown in Table 11.2. The table also includes the percentage variation of the allocation to each sawmill with respect to the direct costs they perceive

Table 11.2 Cost allocations (in kSEK) obtained by different methods and its variation with respect to the direct costs in absence of collaboration with the external company E

Sawmill	K_j	Shapley	Var. (%)	Nucleolus	Var. (%)	EPMm	Var. (%)
N_1	4,057	3,595	−11.4	3,714	−8.5	3,875	−4.5
N_2	2,377	1,803	−24.1	2,965	24.7	2,311	−2.8
N_3	2,073	2,135	3.0	2,180	5.2	2,015	−2.8
N_4	43	5	−87.8	−76	−278.0	41	−2.8
N_5	1,251	1,138	−9.1	1,142	−8.7	1,195	−4.5
N_6	699	660	−5.5	596	−14.7	680	−2.8
N_7	1,670	1,593	−4.6	1,564	−6.3	1,623	−2.8
N_8	678	1,355	99.7	273	−59.8	660	−2.8
N_9	300	325	8.2	280	−6.7	292	−2.8
N_{10}	101	94	−7.7	31	−69.3	99	−2.8
N_{11}	2,004	1,989	−0.7	2,019	0.8	1,948	−2.8
N_{12}	1,715	1,713	−0.1	1,715	0.0	1,667	−2.8
Total	**16,968**	**16,405**	**−3.3**	**16,405**	**−3.3**	**16,405**	**−3.3**

without collaborating with E. All these allocations belong to the core (in the case of the nucleolus and the EPMm this is given by definition and the non-empty core of the game, while for the Shapley values we have verified it numerically). However, we observe notable differences in the results of the sawmills relative to their non-collaborative direct costs. Both Shapley and nucleolus leave some of the sawmills with considerable differences. For example the Shapley allocation for N_8 almost doubles its direct cost, while for N_4 it is equivalent to about 12% of it. N_4 is even more favoured by the nucleolus, which allocates a negative cost to it. The EPMm remedies these issues, by providing a stable allocation that re-distribute the cost as evenly as possible relative to the non-collaborative direct costs of the sawmills. It can be seen in the last column of Table 11.2 that all sawmills get a share of the savings. For most of them, the savings correspond to 2.8% of their non-collaborative direct costs, while for N_1 and N_5 to 4.5%. This situation is a more acceptable solution from a practical point of view in our case, as it reconciles the interest of the personnel responsible for each sawmill with the interest of the whole organization.

11.5 Concluding Remarks

We have studied a problem of re-allocating the cost within an organisation where the units are their own result centers. This is a frequently occurring problem which does not impact the overall organisation but is important for internal purposes such as follow ups and incentive programs. When such an organisation collaborate with an external organisation there are often savings possible. However, for the sub-units the direct cost which impact their result, may differ considerably. This is very clear in our case study.

By applying some standard cost allocations approaches, we can show that the cost allocation results in large variations in the re-allocations. It is easy to motivate that some are unfair and unstable. We proposed a modified EPM method where the associated linear programming model performs the allocation based on the relative difference against the direct cost instead of the stand alone cost which is the standard. This results in a stable allocation of the savings where the difference in relative savings are kept small.

In our results, we have generated the direct cost as the optimal solution to an optimization problem. In practice, other transports were done and paid for. In order to make the actual re-allocation between the companies, there is a need to scale the actual cost with the optimized costs used in the case study.

In our case, we have defined each of the 12 sawmills as independent result centers. A consideration that may be interesting to study further is the fact that there are several owners of the organisation and each of the sawmills belongs to one of these. To balance also on upper level in a two phase allocation could be interesting. An additional consideration is that there may be different alternatives in how to define the direct cost. In our case, we use the optimal allocation to each sawmill when no collaboration is done as the basis. It is also possible to make a re-allocation of the direct cost within the organisation to compensate for events that happened during the execution. For example, the expected supply or demand have changed and this has resulted in urgent re-planning resulting in the use of other supply areas not planned earlier. Such a re-allocation should consider both the planned transportation as well as the operated transportation. Finally, while our focus has been on cost savings, allocating emissions among the partners could be an alternative approach in response to nowadays environmental concerns.

References

Audy J-F, Di Amours S, Rousseau L-M (2011) Cost allocation in the establishment of a collaborative transportation agreement an application in the furniture industry. J Oper Res Soc 62(6):960–970

Flisberg P, Frisk M, Rönnqvist M, Guajardo M (2015) Potential savings and cost allocations for forest fuel transportation in Sweden: a country-wide study. Energy 85:353–365

Flisberg P, Frisk M, Guajardo M, Rönnqvist M (2016) The aftermath of implementing collaboration in a network of sawmills: a retrospective analysis on logistics costs. In: ILS Conference Information Systems, Logistics and Supply Chain, Bordeaux, France, June 1–4, 2016. http://ils2016conference.com/wp-content/uploads/2015/03/ILS2016_TE04_2.pdf

Forsberg M, Frisk M, Rönnqvist M (2005) FlowOpt-a decision support tool for strategic and tactical transportation planning in forestry. Int J For Eng 16(2):101–114

Frisk M, Göthe-Lundgren M, Jörnsten K, Rönnqvist M (2010) Cost allocation in collaborative forest transportation. Euro J Oper Res 205(2):448–458

Fromen B (1997) Reducing the number of linear programs needed for solving the nucleolus problem of n-person game theory. Euro J Oper Res 98(3):626–636

Guajardo M, Jörnsten K (2015) Common mistakes in computing the nucleolus. Euro J Oper Res 241(3):931–935

Guajardo M, Rönnqvist M (2015) Operations research models for coalition structure in collaborative logistics. Euro J Oper Res 240(1):147–159

Guajardo M, Rönnqvist M (2016) A review on cost allocation methods in collaborative transportation. Int Trans Oper Res 23(3):371–392

Guajardo M, Jörnsten K, Rönnqvist M (2016) Constructive and blocking power in collaborative transportation. OR Spectr 38(1):25–50

Hitchcock FL (1941) The distribution of a product from several sources to numerous localities. J Math Phys 20(1):224–230

Schmeidler D (1969) The nucleolus of a characteristic function game. SIAM J Appl Math 17(6):1163–1170

Shapley LS (1953) A values for n-person games. In: Kuhn HW, Tucker AW (eds) Contributions to the theory of games, vol II (Annal of Mathematics Studies)

Temponi C, Vandaele N (eds) LNBIP 262: ILS 2016. Springer (under review)

Tolstoi AN (1930) Methods of finding the minimal total kilometrage in cargo transportation planning in space. TransPress of the National Commissariat of Transportation, pp 23–55

Chapter 12
Impact of the Heterogeneity of the Ballast on the Dynamical Behavior of the Ballast-Soil System

Lucio De Abreu Correa, Regis Cottereau, Estelle Bongini, Sofia Costa d'Aguiar, Baldrik Faure and Charles Voivret

Abstract This paper discusses the dynamical behavior of a randomly-fluctuating heterogeneous continuum model of the ballast. The Young's modulus is modeled as a random field parameterized by its average, its variance and a correlation model representing non-interpenetrating spheres. A numerical model of the ballast and the surrounding soil is then constructed based on an efficient implementation of an explicit spectral element solver on a large cluster of computers. This model allows to describe numerically the wave field generated in the ballast and soil by the passage of a train, as well as to construct dispersion equations for the ballast-soil model. The influence of heterogeneity is discussed by comparison with a similar model where the ballast is assumed homogeneous. Different values of the soil mechanical parameters are considered and compared. Finally, potential consequences for the design of the ballast are discussed.

L. De Abreu Correa · R. Cottereau (✉)
Laboratoire MSSMat UMR 8579, CentraleSupelec, CNRS, Châtenay-Malabry, France
e-mail: regis.cottereau@centralesupelec.fr

L. De Abreu Correa
e-mail: lucio.de-abreu-correa@ecp.fr

E. Bongini · S. Costa d'Aguiar · B. Faure · C. Voivret
SNCF, Direction de l'Innovation et de la Recherche, Paris, France
e-mail: estelle.bongini@sncf.fr

S. Costa d'Aguiar
e-mail: sofia.costadaguiar@sncf.fr

B. Faure
e-mail: Baldrik.faure@sncf.fr

C. Voivret
e-mail: charles.voivret@sncf.fr

P. Diez et al. (eds.), *Computational Methods and Models for Transport*,
Computational Methods in Applied Sciences 45,
DOI 10.1007/978-3-319-54490-8_12

12.1 Introduction

Although the annual funding for railway tracks construction and maintenance is high (the public funding exceeds $2.1 billion per year, only in Australia (BITRE 2003), a large part of track design and maintenance planning is still based on strongly empirical grounds. The passage of trains is the main source of damage on the track and hence contributes heavily to this budget. In the past years, with the increasing average velocity of the train passages, dynamical effects have been shown to accelerate track deterioration and cause heavier damage to neighbor buildings (Connolly et al. 2014). Understanding precisely the dynamical behavior of ballasted railway tracks is therefore paramount to proposing improved design strategies and maintenance programs.

Experimental work on small granular samples indicates that the heterogeneity of the material constitutes the main reason for its peculiar dynamical behavior. In particular, it has been shown that the wave propagation velocity depends on the confining pressure through the complex assemblage of contacts (Goddard 1990). In experiments of ultrasound wave propagation in granular media, the appearance of an incoherent coda behind the main pulse is also due to the heterogeneity of the medium (Jia et al. 1999). Although at a different scale, the latter observation is classical in seismic wave propagation and has also been attributed to the heterogeneity of the Earth crust (Aki 1969).

Analytical models have been constructed to try and explain the influence of heterogeneity on the dynamical behavior of granular material. For instance, (Leibig 1994) used springs with random properties to reproduce the complex frequency response seen in measurements of sound propagation. More recently, similar frequency filtering results were revisited in Lawney and Luding (2014). In the context of ballasted railway tracks, attempts have been made also at using beams lying over winkler-type springs (Suiker 2002; Sheng et al. 2004). However, more realistic models are called for to discuss actual engineering systems. Two classes of numerical models are used to try and predict the behavior of these dynamical systems: (1) discrete approaches, in which each grain of the ballast is represented by a rigid body and interacts with its neighbors through nonlinear contact forces (using molecular dynamics or non-smooth contact dynamics); and (2) continuum approaches, in which the ballast is replaced by a homogenized continuum and a Finite Element Method (FEM) is used to solve the problem. Although they are very efficient at representing complex nonlinear behaviors (Drescher and Josselin Jong 1972; Liu et al. 1995; Bagi 2003), the discrete approaches are today capable of solving only a few meters-length of ballast. The development of an efficient scheme for parallelization over large clusters of computers and the coupling with continuum models remains an open problem (Thi Minh Phuong et al. 2011; Wellmann and Wriggers 2012). On the other hand, FEM (in particular 2.5D FEM (Freitas da Cunha and José Pedro 2013; Arlaud et al. 2014; Bian et al. 2015) and semi-analytical FEM (Lagasse 1973)) and Boundary Element methods (Freitas da Cunha and José Pedro 2013), all based on continuum models, have been used to compute the dynamical behavior of ballasted railway tracks over large distances. Unfortunately, these models are all based on an assumption of homogeneity of the ballast layer. They hence cannot reproduce the typical behavior of granular materials.

This paper discusses the dynamical behavior of a recently proposed randomly-fluctuating heterogeneous continuum model of the ballast (Correa et al. 2016b). The Young's modulus is modeled as a random field parameterized by its average, its variance and a correlation model representing non-interpenetrating spheres. Being continuous, a numerical model of the ballast and the surrounding soil can then be constructed based on an efficient implementation of an explicit spectral element solver on a large cluster of computers. Being heterogeneous, it is expected that the complex behavior of granular media can be better reproduced than with previous homogeneous continuum models. Our model allows to describe numerically the wave field generated in the ballast and soil by the passage of a train (see Sect. 12.2), as well as to construct dispersion equations for the ballast-soil model (see Sect. 12.3). The influence of heterogeneity is discussed by comparison with a similar model where the ballast is assumed homogeneous. Different values of the soil mechanical parameters are considered and compared. In particular, we compare the case of a soil with the same properties as the ballast to that of a ballast resting over a stiff soil. Finally, potential consequences for the design of the ballast are hinted at in conclusion.

12.2 Influence of the Heterogeneity of the Ballast Layer on the Wave Field Induced by the Passage of a Train

The objective of this section is to discuss the wave field induced in the ballast and in the soil by the passage of a train. In particular, we will concentrate on the influence of the heterogeneity of the ballast layer on that wave field, by comparing simulation results for homogeneous and heterogeneous ballast layers. Section 12.2.1 introduces the general setting of the numerical models, valid for both the homogeneous and heterogeneous ballast layers. Section 12.2.2 reminds the main ingredients of the randomly-fluctuating model of Young's modulus, used to model the heterogeneous ballast, and described in more details in Correa et al. (2016b). Section 12.2.3 describes the Spectral Element solver used to simulate the wave fields. Finally, Sect. 12.2.4 describes the general wave field induced by the passage of a train on a ballasted railway track and compares the cases with homogeneous and heterogeneous ballast layers.

12.2.1 General Setting of the Numerical Models

We consider the model depicted in Fig. 12.1. It is a 38 m-long one-way track segment, with a ballast (in yellow on Fig. 12.1) of height 48 cm and width between 3.9 m at the top and 5 m at the bottom. Below the track, the soil is numerically modeled on a width of 20 m and a depth of 5 m. The concrete sleepers have dimensions

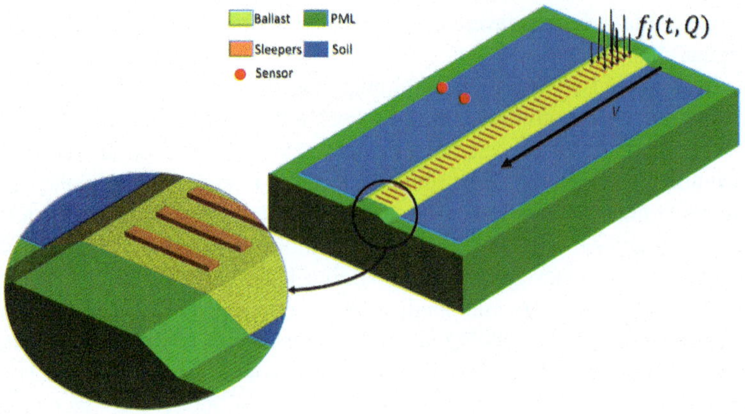

Fig. 12.1 Geometry, loading, boundary conditions and mesh for the wave propagation analysis

$20 \times 30 \times 200\,\text{cm}^3$, are separated by $d = 0.6\,\text{m}$, and are embedded in the soil on a height of 10 cm. In the simulated domain, the wave field is the solution of the wave equation:

$$\nabla \cdot \sigma - \rho \ddot{\mathbf{u}} = \mathbf{0} \tag{12.1}$$

where the material is assumed isotropic and linear, so that the strain is $\epsilon = (\nabla \mathbf{u} + \nabla \mathbf{u}^T)/2$ and the stress is $\sigma = \lambda(\mathbf{x})\text{Tr}\epsilon \mathbb{I} + 2\mu\epsilon$, with λ and μ the Lame parameters, and $\mathbb{I}$ is the identity second-order tensor. The density is denoted ρ. The Lame parameters are related to the wave velocities through $V_p = \sqrt{(\lambda + 2\mu)/\rho}$ and $V_s = \sqrt{\mu/\rho}$. The soil is assumed homogeneous in all simulations, with $V_s = 180\,\text{m/s}$, $V_p = 350\,\text{m/s}$, and $\rho = 1900\ \text{kg/m}^3$. The sleepers are assumed homogeneous in all simulations, with $V_s = 2500\,\text{m/s}$, $V_p = 4500\,\text{m/s}$, and $\rho = 2400\,\text{kg/m}^3$. In the simulations where the ballast is assumed homogeneous, we consider $V_s = 150\,\text{m/s}$, $V_p = 380\,\text{m/s}$, and $\rho = 1900\,\text{kg/m}^3$. In the simulations where the ballast is assumed heterogeneous, the random model of mechanical properties is described in next section, and the average values are taken equal to those of the homogeneous case.

The loading considered is a classical one for modeling the influence of a train-rail system on the sleepers (Guerin 1996). It considers the flexibility of the rail by transferring the point loads of the boggie of a train onto consecutive sleepers. Although horizontal components could be added (in the case of a curved track for instance), we consider here only vertical loading. The movement of the train is taken into account by moving the position of the point loads at the appropriate velocity. The loading on each of the sleepers (and on each side of each sleepers, as indicated on Fig. 12.1) is given by:

$$F_i(t) = \frac{QY}{2}\left[C^{\frac{(v_0(t-\delta_i)-a)^2}{d^2}} + C^{\frac{(v_0(t-\delta_i)-a-L)^2}{d^2}}\right] \tag{12.2}$$

where Q is the load magnitude, $L = 3\,\text{m}$ is the wheelbase, $a = 5d = 3$ m is the critical distance, beyond which the load is assumed to vanish, $v_0 = 100\,\text{m/s}$ is the chosen train velocity, and $C = 0.61$ and $Y = 0.41$ are constants that depend of the soil-ballast combination. The latter values are obtained from the experimental values in Al Shaer et al. (2008). In order to produce a "moving load" each sleeper is associated with a delay δ_i. The Fig. 12.1 highlights the region where the vertical load is applied at the top of the sleepers at the initial time. Only five sleepers are loaded in the image because the influence of the boggie on further sleepers almost vanishes. The sensors used to record the displacement field in the model are also represented in the same figure.

In the inset of Fig. 12.1, the refinement of the mesh is also shown. The mesh is discretized with 0.81 millions of hexahedral elements. High-order polynomials of order 7 are used in the spectral element solver (see Sect. 12.2.3), which corresponds to 343 degrees of freedoms (DOFs) for each element, and a total of close to 175 millions of DOFs for the entire mesh. On the exterior of the soil box (green area in Fig. 12.1), a Perfectly Matched Layer (Festa and Vilotte 2005) is added to absorb outgoing waves.

12.2.2 Randomly-Fluctuating Heterogeneous Continuum Model of the Ballast

The objective of the paper is to investigate the influence of heterogeneity of the ballast on the wave field. We describe in this section the randomly-fluctuating heterogeneous continuum model that is considered here for the ballast. It is described in detail in (Correa et al. 2016b) and only the general principles are recalled here. Young's modulus E is the only parameter that is assumed to be heterogeneous, although more complex media could be considered, including anisotropic (Ta et al. 2010) or non-linear (Jehel and Cottereau 2015). The correlation model of Young's modulus is constructed analytically and corresponds to that of a dense packing of spheres with known diameter and packing density (both quantities measurable on real ballast), as plotted in Fig. 12.2. The first-order marginal law is obtained by identification on Discrete Element simulations of granular samples. A gamma law is considered, with average $\mathbb{E}[E] = 95$ MPa and variance $\mathbb{E}[(E - \mathbb{E}[E])^2] = 135 \times 10^3(\text{MPa})^2$. Note in particular that the coefficient of variation is extremely high, on the order of several times the average. As desired, the model therefore recreates the heterogeneity of the granular material, that has been shown to be influential on its dynamical behavior. However it does not capture its discrete character, and hence cannot take into account rearrangements or changes of topology that would occur in real granular samples.

The generation of realizations of Gaussian random field is performed using the spectral representation method, as described originally in Shinozuka and Deodatis (1991). This method consists in summing a large number of cosine functions with amplitudes chosen to match the desired correlation model and random phases:

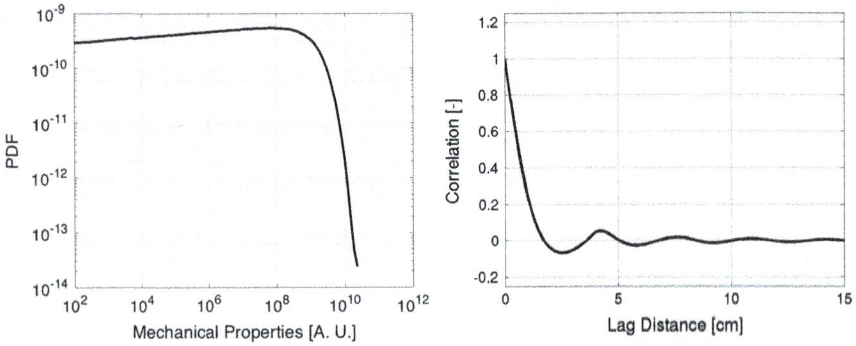

Fig. 12.2 First-order marginal law (*left* figure) and correlation model (*right* figure) of the ballast

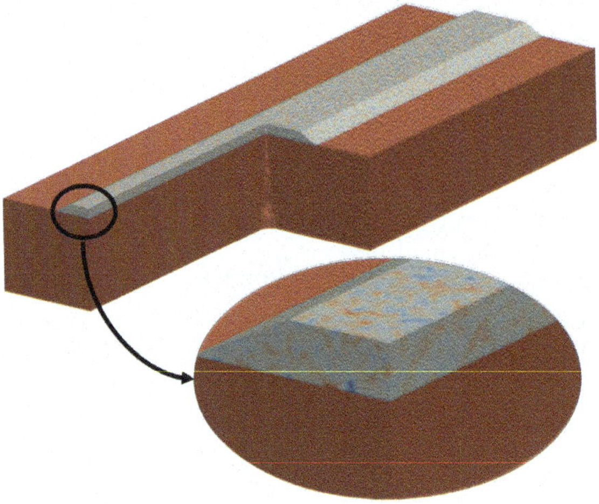

Fig. 12.3 One realization of Young's modulus of the soil and heterogeneous ballast

$$E(\boldsymbol{x}) = 2\sum_{j=1}^{N}\sqrt{|\Delta\mathbf{k}|\mathrm{S}(\mathbf{k}_j)}\cos(\mathbf{k}_j\cdot\boldsymbol{x}+\phi_j), \tag{12.3}$$

where $\mathrm{S}(\mathbf{k})$ is the Fourier transform of the correlation and the ϕ_j are random germs with uniform distribution over $[0, 2\pi]$. The efficiency of the generation can be improved by considering a Fast Fourier Transform algorithm, and we consider a particular implementation that allows to generate realizations over very large clusters of processors. This is obtained by localizing the generation on each processor before merging at the interface to retrieve the desired correlation properties (Paludo et al. 2016). Once a realization of a centered Gaussian field has been obtained, it is then transformed into the desired first order marginal law by a non-linear mapping (Rosenblatt 1952). Figure 12.3 presents one realization of the random field of Young's modulus for the ballast-soil system.

12.2.3 Spectral Element Solver

To approximate the solution of the wave equation, the solver used in this paper is based on the Spectral Element Method (SEM, see Cohen (2001) for mathematical details and Komatitsch (2005) for an example in seismology). The SEM is a high-order FEM that uses a Gauss-Lobatto-Legendre quadrature rule for integration and Lagrange polynomials based on the nodes of that quadrature. Inserting the polynomial functions and quadrature rules into the variational form of Eq. (12.1) leads to a system of ordinary differential equations:

$$\mathbf{M}\dot{\mathbf{V}} = \mathbf{F}_{\mathbf{ext}} - \mathbf{F}_{\mathbf{int}}(\mathbf{U}) \tag{12.4}$$

where $\mathbf{U}$ and $\mathbf{V}$ are vectors containing the components of the displacement and velocity at the nodes, respectively, $\mathbf{M}$ is the mass matrix, and the vectors $\mathbf{F}_{\mathbf{ext}}$ and $\mathbf{F}_{\mathbf{int}}$ contain the influence of the passing train and the internal forces, respectively. As it uses high-order polynomials, the method is exponentially accurate: an increase of the polynomial order leads to an exponential decrease of the error. This property, called spectral precision, gives its name to the method. Using under-integration, the mass matrix becomes naturally diagonal, which allows to use an explicit second-order finite-difference scheme in time. Even though the stability condition requires to use very small time steps, the construction of the solution at each time step is very easy because the inversion of the mass matrix is instantaneous. This feature does not occur in classical high-order FEM. Our implementation of the SEM has demonstrated scalability for more than 10’000 cores (Guillot et al. 2014) while other implementations in the literature have even been shown to scale over 100’000 cores (Komatitsch 2005). Note that our implementation is capable of dealing with externally-created 3D unstructured meshes, as is the case for the ballast.

12.2.4 Influence of Heterogeneity on the Wave Field

This section discusses the wave fields induced in the ballast-soil system by the passage of a train, using the numerical method of Sect. 12.2.3 and models of Sects. 12.2.1–12.2.2. As explained earlier, we consider two different models, and compare them. The models are similar in all features except for the mechanical parameters in the ballast layer: one of the model has homogeneous properties in the ballast, while the other is heterogeneous. Note that the average value of the heterogeneous Young’s modulus is equal to the value of the homogeneous Young’s modulus. Note also that the input force is normalized *a posteriori* in order to make sure that the total energy introduced in the two models is the same. The displacement field at time $t = 0.27\,\text{s}$ in the case when the ballast is homogeneous is plotted at the left of Fig. 12.4. The pattern is quite simple, with an energy clearly concentrated under the moving load. Most of the displacement seems to be concentrated in the ballast layer, with rapid decrease both in depth and at distance from the track. There seems

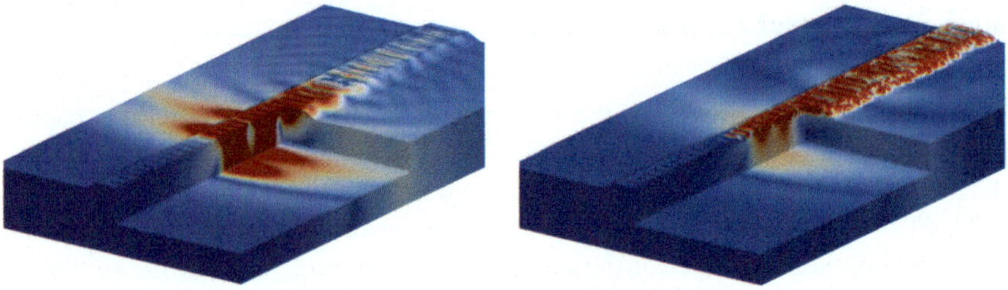

Fig. 12.4 Displacement fields at time $t = 0.27$ s. The *left* figure corresponds to the simulation with homogeneous ballast, while the *right* one corresponds to simulation with heterogeneous ballast

to be mainly a guided wave within the ballast layer, connecting with Rayleigh wave in the soil. At the right of Fig. 12.4, the same displacement field is plotted in the case of the heterogeneous ballast. The wave pattern is very different, with most of the energy remaining within the ballast, and not necessarily concentrated below the moving load. The energy seems to be trapped within the ballast layer, but not moving forward as in the case of guided waves. Overall, the energy radiated in the soil seems to be smaller.

This last statement is confirmed by the times histories at the sensors marked in Fig. 12.1 and plotted in Fig. 12.5. Indeed, the maximum values of the displacement field are larger with a homogeneous ballast than with a heterogeneous one. On the other hand, after some time, the tendency is inverted. As most of the energy in the

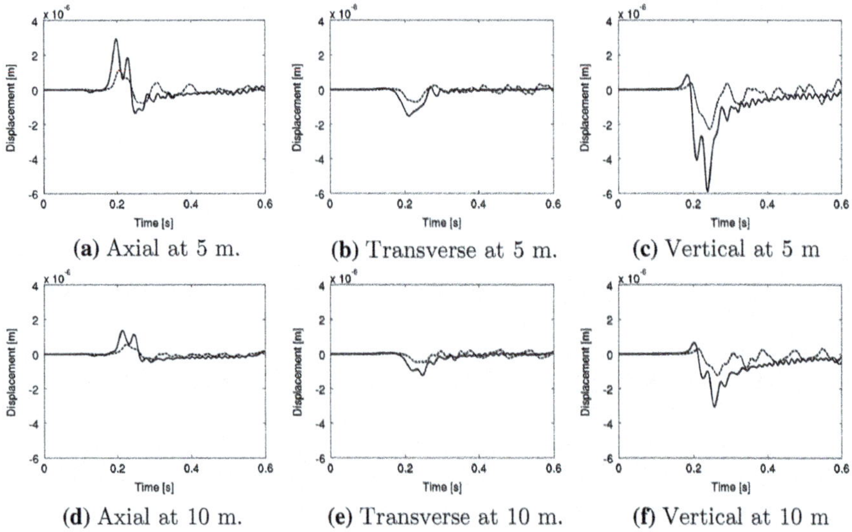

(**a**) Axial at 5 m. (**b**) Transverse at 5 m. (**c**) Vertical at 5 m

(**d**) Axial at 10 m. (**e**) Transverse at 10 m. (**f**) Vertical at 10 m

Fig. 12.5 Displacements at the two sensors, located at 5 and 10 m from the track (see Fig. 12.1 for the exact position). The solid lines correspond to the simulation with homogeneous ballast, while the dashed ones correspond to simulation with heterogeneous ballast

homogeneous case remains concentrated on a single front, once that front has passed, there is little energy remaining in the system. For a heterogeneous ballast, the energy initially trapped in the ballast keeps on sending waves into the soil for a longer time. Note that there is no dissipation in our model, so that the later part of the recordings are probably not very realistic. Within an actual ballast, a large part of the energy trapped in the ballast layer would be dissipated by friction and crushing of grains. In all cases, there seems to be apparent damping due to the heterogeneity in the ballast.

12.3 Dispersion Curves in a Continuum Heterogeneous Ballast

We now turn to a description of the same physical situation in terms in dispersion curves. Assuming that the displacement field in Eq. (12.1) is expressed as $\mathbf{u} = \tilde{\mathbf{u}} \exp i(\mathbf{k} \cdot \mathbf{x} - \omega t)$, and simplifying by the density, we obtain:

$$(\Gamma(\mathbf{k}) - \omega^2 \mathbb{I})\tilde{\mathbf{u}} = \mathbf{0} \tag{12.5}$$

where $\rho\Gamma(\mathbf{k}) = (\lambda + \mu)\mathbf{k} \otimes \mathbf{k} - i(\nabla\lambda \otimes \mathbf{k} + \mathbf{k} \otimes \nabla\mu) + (\mu|\mathbf{k}|^2 - i\mathbf{k} \cdot \nabla\mu)$, $\mathbf{k}$ is the wave number, ω is the circular frequency, and $\Gamma(\mathbf{k})$ is Christoffel tensor. The dispersion equations are defined as $\det(\Gamma(\mathbf{k}) - \omega^2 \mathbb{I}) = 0$, and indicate the $(\mathbf{k}, \omega)$ pairs that allow the existence of non-vanishing solutions of the wave equation. Except for very simple geometries and boundary conditions (Lagasse 1973; Knopoff 1964), this dispersion equation cannot be constructed analytically. We therefore provide in this section a numerical construction of the dispersion equation for a heterogeneous ballast. Section 12.3.1 describes the general method to obtain the dispersion equation based on solutions of the wave equation in the time×space domain. Section 12.3.2 then describes the two numerical models that are considered in this paper, with a stiffer or a softer underlying soil, each with a homogeneous counterpart for comparison. The following section addresses the wave patterns obtained for the different models. Finally, Sect. 12.3.4 describes the dispersion relations for the ballast.

12.3.1 General Methodology to Obtain Dispersion Curves Using Time-Space data

The general methodology to construct the dispersion relation consists in computing a displacement field $\mathbf{u}(\mathbf{x}, t)$, and performing Fourier transforms in both the time and space variables to obtain $\tilde{\mathbf{u}}(\mathbf{u}, \omega)$:

$$\tilde{\mathbf{u}}(\mathbf{u}, \omega) = \int_{-\infty}^{+\infty} \int_{-\infty}^{+\infty} \mathbf{u}(\mathbf{x}, t) e^{-i(\mathbf{k} \cdot \mathbf{x} + \omega t)} d\mathbf{x} dt \tag{12.6}$$

The locations in $(\mathbf{k}, \omega)$ where the amplitude $\|\tilde{\mathbf{u}}(\mathbf{u}, \omega)\|$ is large match the dispersion equation (Alleyne and Cawley 1991).

In principle, the dispersion equation should be constructed in the 4-dimensional space $(\mathbf{k}, \omega)$. However, as we are mainly interested in the propagation of waves along the axis of the ballast, we will rather consider a partial Fourier transform:

$$\tilde{\mathbf{u}}_x(k_x, \omega; y_i, z_i) = \int_{-\infty}^{+\infty} \int_{-\infty}^{+\infty} \mathbf{u}(\mathbf{x}, t) e^{-i(k_x x + \omega t)} dx dt \tag{12.7}$$

where x is the variable along the axis of the ballast and y_i and z_i are particular positions within the section of the ballast. We therefore concentrate on modes propagating along the axis of the ballast. As the amplitude $\|\tilde{\mathbf{u}}_x(k_x, \omega; y, z)\|$ depends on the position within the section (y_i, z_i), we will consider three different heights within the ballast: top ($z_t = 48$ cm), middle ($z_m = 24$ cm) and bottom ($z_b = 0$ cm), the latter being at the interface between the ballast and the underlying soil. Finally, we will only consider positions in the symmetry plane of the ballast ($y_0 = 0$ cm), although other positions might be interesting (in particular for higher frequencies when the energy concentrates in the shoulders of the ballast layer). Hence, we will construct for each numerical model three two-dimensional dispersion curves: $\|\tilde{\mathbf{u}}_x(k_x, \omega; y_0, z_t)\|$, $\|\tilde{\mathbf{u}}_x(k_x, \omega; y_0, z_m)\|$ and $\|\tilde{\mathbf{u}}_x(k_x, \omega; y_0, z_b)\|$.

12.3.2 *Description of the Numerical Models*

As mentioned above, we will construct dispersion curves for two different models: one with a stiffer ballast, and one with a softer ballast. In each case, we will construct a homogeneous model, with homogeneous ballast and soil, and a heterogeneous one, with heterogeneous ballast and homogeneous soil.

12.3.2.1 Frequency-Range Dependent Meshes

As we aim at creating dispersion curves for a large frequency band [8–400] Hz, three different models are created for each case of interest, centered and constructed for different frequency bands [8–60] Hz, [25–200] Hz, and [55–400] Hz. More specifically, Ricker pulses (second derivative of a Gaussian) with unit amplitude and central frequencies of respectively 30, 100 and 210 Hz are considered. Because larger models are required to represent long wavelengths, while smaller elements are necessary for small wavelengths, three different numerical meshes are used for the three excitations. Their characteristics are summarized in Table 12.1.

As an example, the High-Frequency mesh is shown in Fig. 12.6. It is 32 m-long, and is discretized with 0.25 millions of elements, with 375 degrees of freedoms (DOFs) in each element, which amounts to a total of $\approx$ 94 millions DOFs in the entire model. The same figure also highlights the region where the vertical load is

Table 12.1 Numerical parameters for the models constructed for each frequency band

	Low-frequency	Middle-frequency	High-frequency
Central frequency [Hz]	30	100	210
Length [m]	48	40	32
Characteristic Elt. size [m]	0.05	0.05	0.05
Distance between sensors [m]	0.25	0.05	0.05
Δt [s]	$1.75\ 10^{-6}$	$1.75\ 10^{-6}$	$1.75\ 10^{-6}$
Number of elements	380000	320000	250000
Total DOFs [-]	$\approx 132 \times 10^6$	$\approx 120 \times 10^6$	$\approx 94 \times 10^6$
CPU time [h]	3800	2500	1700

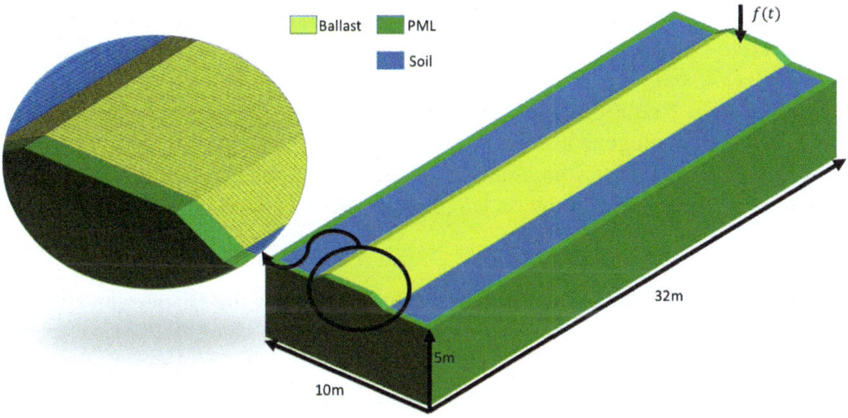

Fig. 12.6 Geometry, mesh, and loading for the high-frequency case

applied at the top of the ballast layer. Note that this load is not moving, contrarily to the load considered in Sect. 12.2. Note also that the space discretization of the three meshes are appropriate for both the soft and stiff soil models, whose material parameters will be described further.

12.3.2.2 Soft Underlying Soil Models

The first model we consider is a model where the average properties of the ballast are the same as those of the soil. This is not a realistic situation because the underlying soil is usually stiffer than the ballast layer. However, it is interesting to study from a physical point of view because it is expected that the dispersion curves will be simpler in that case. Indeed, at least in the homogeneous case, the ballast will not behave as a guiding medium, and the overall medium should behave similarly as a half-space.

The comparison between the homogeneous and heterogeneous ballast is also expected to be simpler because the influence of heterogeneity will not be polluted by the presence of guided modes. For the homogeneous model, both the soil and ballast have the following mechanical properties: $V_s = 150\,\text{m/s}$, $V_p = 380\,\text{m/s}$, and $\rho = 1900\,\text{kg/m}^3$. For the heterogeneous model, the soil is still assumed homogeneous, and the average of the Young's modulus of the ballast is assumed to be equal to the Young's modulus of the homogeneous case. The variance and correlation length are those described in Sect. 12.2.2.

12.3.2.3 Stiff Underlying Soil Models

For this model, the objective is to construct a dispersion equation including the effects of both the heterogeneity within the ballast and of the mismatch of material properties between the ballast and the underlying soil. For the homogeneous case, we therefore consider the same properties as above for the ballast: $V_s = 150\,\text{m/s}$, $V_p = 380\,\text{m/s}$, and $\rho = 1900\,\text{kg/m}^3$, and the following properties for the soil: $V_s = 180\,\text{m/s}$, $V_p = 1100\,\text{m/s}$, and $\rho = 1900\,\text{kg/m}^3$. For the heterogeneous model, the soil is still assumed homogeneous, and the average of the Young's modulus of the ballast is assumed to be equal to the Young's modulus of the ballast in the homogeneous case. The variance and correlation length are those described in Sect. 12.2.2.

12.3.2.4 Description of the Sensors for Estimation of the Dispersion Curves

As described in Sect. 12.3.1, we aim at constructing three two-dimensional dispersion curves: $\|\tilde{\mathbf{u}}_x(k_x, \omega; y_0, z_t)\|$, $\|\tilde{\mathbf{u}}_x(k_x, \omega; y_0, z_m)\|$ and $\|\tilde{\mathbf{u}}_x(k_x, \omega; y_0, z_b)\|$ for each of the models. We therefore need to measure displacement fields $\mathbf{u}(\mathbf{x}, t)$ along three positions within the section of the ballast: in (y_0, z_t), (y_0, z_m) and (y_0, z_b), with sufficient discretization in the axial direction so as to be able to compute the Fourier transform along x. Figure 12.7 therefore shows the position of the sensors used to record the displacement field in the Low-Frequency mesh.

12.3.3 Analysis of the Wave Patterns

Before describing the dispersion equations, we present below general wave patterns obtained in the models considered. Note that the main difference with the previous case is that the loading is static. The simulations are performed with the spectral element solver described in Sect. 12.2.3.

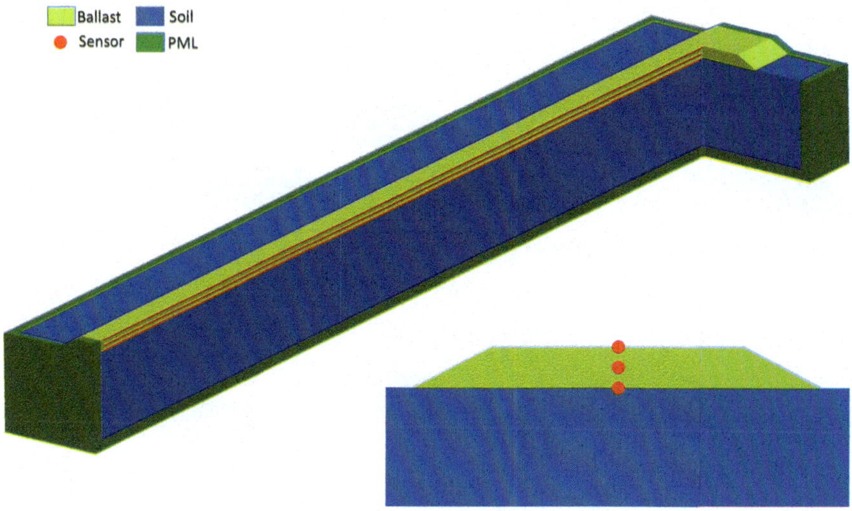

Fig. 12.7 Cut view of the geometry and position of the sensor lines (*red dots*). Three different sets are considered at the *top* of the ballast, in the *middle* of the layer, and at the interface with the soil

12.3.3.1 Simulations with a Soft Underlying Soil

In Fig. 12.8, snapshots are plotted at different times for the three homogeneous models with soft soil. Each line represents a different model, with the upper one corresponding to the lower frequency excitation and the lower one to the higher frequency excitation. From left to right, we observe snapshots at times $t = 0.05$ s, $t = 0.15$ s and $t = 0.25$ s, and clearly see a wave emitted at the right of the railway track propagating to the left. As the soil has the same properties as the ballast, the model is very similar to a half-space and we indeed see a wave propagating essentially as in a half-space, with very little influence of the shoulders of the track. The main difference between the three excitation frequencies is that the propagating wave has shorter wavelength.

We now consider the simulations in a heterogeneous model, where the soil is homogeneous and the ballast is heterogeneous. They are plotted in Fig. 12.9, and several comments can be made. At lower frequencies, the energy still propagates, but there is a clear concentration of the amplitudes within the ballast layer. Although the average values of the mechanical parameters in the ballast are the same as the values in the soil, the overall effect of the heterogeneity is to create a mismatch that retains the energy within the ballast. At higher frequencies, this effect increases and is completed by the creation of a strong coda. This coda is seen as a transformation of the coherent energy in the main pulse into a trailing incoherent energy widely dispersed behind that pulse. Finally at the largest frequency, part of the coda is not even propagating, remaining right below the input energy. This phenomenon is called

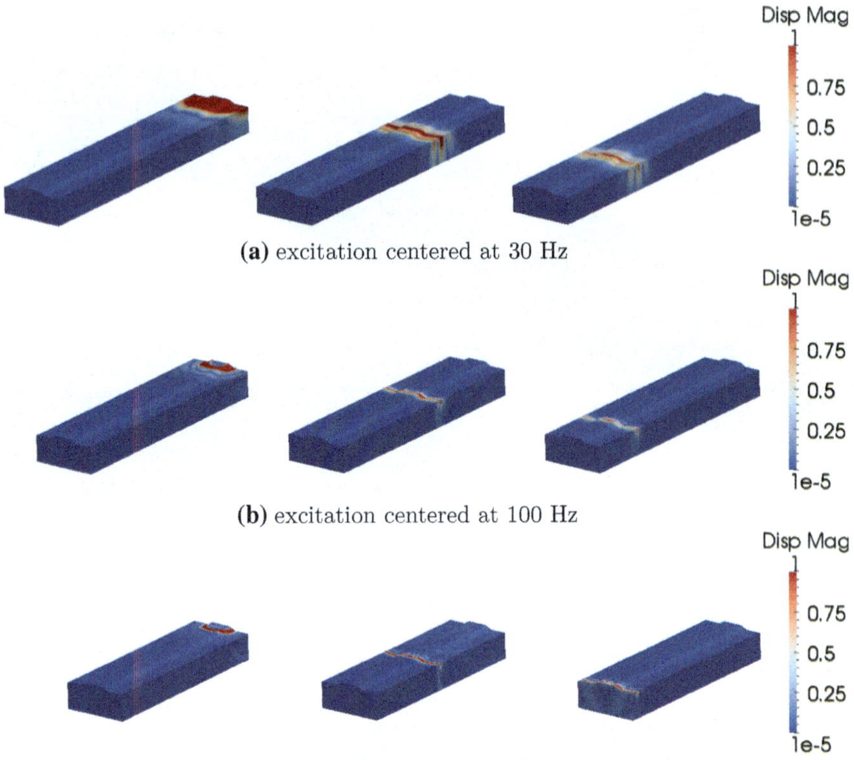

(a) excitation centered at 30 Hz

(b) excitation centered at 100 Hz

(c) excitation centered at 210 Hz

Fig. 12.8 Displacement fields in the homogeneous model with soft soil at times $t = 0.05$ s (*left column*), $t = 0.15$ s (*center column*) and $t = 0.25$ s (*right column*). The different models are centered around frequencies 30 Hz (*upper line*), 100 Hz (*central line*) and 210 Hz (*lower line*)

strong localization (Lawney and Luding 2014), and is a direct consequence of the random heterogeneity of the ballast properties.

12.3.3.2 Simulations with a Stiff Underlying Soil

We now turn to simulations where the soil is stiffer than the ballast, as expected in real applications, and start with the homogeneous case, where both soil and ballast are homogeneous, and plotted in Fig. 12.10. For the lowest frequency, there is little difference with the case where the properties of the soil and the ballast were the same. As the wavelength is larger than the thickness of the ballast, homogenization takes places and the behavior is similar to that of a homogeneous model. For higher frequencies, the guiding behavior comes into play and the energy seems to be trapped within the ballast. Note however that the physical origin of the trapping is completely

(a) excitation centered at 30 Hz

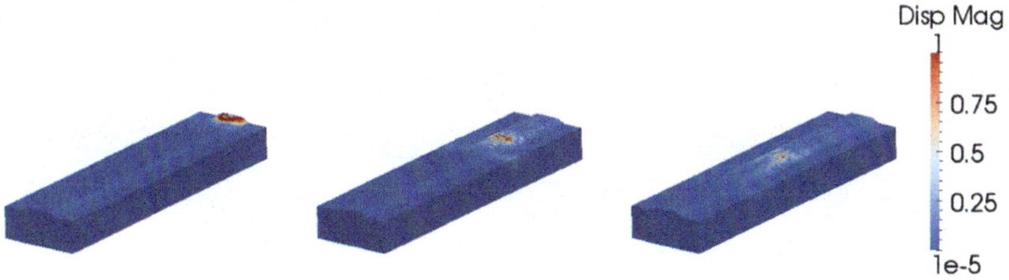

(b) excitation centered at 100 Hz

(c) excitation centered at 210 Hz

Fig. 12.9 Displacement fields in the heterogeneous model with soft soil at times $t = 0.05\,\mathrm{s}$ (*left column*), $t = 0.15\,\mathrm{s}$ (*center column*) and $t = 0.25\,\mathrm{s}$ (*right column*). The different models are centered around frequencies 30 Hz (*upper line*), 100 Hz (*central line*) and 210 Hz (*lower line*)

different as above, so that the energy remains propagating even at higher frequencies. We also observe amplification at the shoulders of the ballast that are typical of waveguides with that shape.

Finally, in Fig. 12.11, we observe the case of a heterogeneous ballast over a stiff homogeneous soil, which is the closest to a real ballasted railway track. The overall behavior is a combination of those described above. At low frequency, we observe a main pulse propagating rather unhindered by the heterogeneity of the ballast or the impedance mismatch between the ballast and the soil. Behind that coherent pulse, a coda is observed, propagating slowly within the ballast layer. At higher frequencies, the coherent pulse completely disappears, transferring its energy partly into a slowly-propagating coda and a stationary localized state just below the excitation.

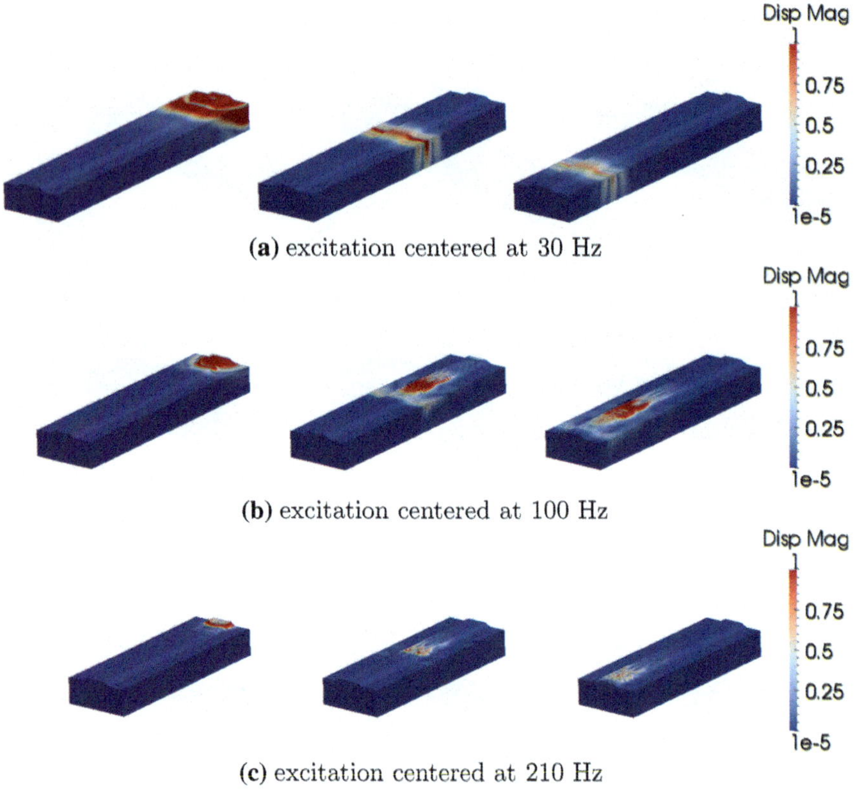

Fig. 12.10 Displacement fields in the homogeneous model with stiff soil at times $t = 0.05$ s (*left column*), $t = 0.15$ s (*center column*) and $t = 0.25$ s (*right column*). The different models are centered around frequencies 30 Hz (*upper line*), 100 Hz (*central line*) and 210 Hz (*lower line*)

12.3.4 Dispersion Equation

We finally turn to the construction of the dispersion curves for the different models, based on (see Sect. 12.3.1) Fourier transforms of the measurements acquired (sum of all displacement components) along the sensors lines indicated in Fig. 12.7. Note that the figures that are presented here are obtained by summing the contributions from the simulations with different loading frequencies. Note also the Fourier transforms are all normalized with the Fourier transform or their respective loading.

12.3.4.1 Dispersion Curves for the Ballast Resting on Soft Soil

We start with the homogeneous model of the ballast, with properties matching those of the underlying soil. The dispersion curves are plotted in Fig. 12.12, and correspond

(**a**) excitation centered at 30 Hz

(**b**) excitation centered at 100 Hz

(**c**) excitation centered at 210 Hz

Fig. 12.11 Displacement fields in the heterogeneous model with stiff soil at times $t = 0.05$ s (*left column*), $t = 0.15$ s (*center column*) and $t = 0.25$ s (*right column*). The different models are centered around frequencies 30 Hz (*upper line*), 100 Hz (*central line*) and 210 Hz (*lower line*)

to the snapshots of Fig. 12.8. As expected from the snapshots, we retrieve approximately the dispersion relation for a Rayleigh wave in a homogeneous medium, with a velocity close to that of a shear wave (indicated by the black dotted line on Fig. 12.12). Although most of the displacement field corresponds to shear along the line that we are measuring, a small amount of P wave is also encountered at extremely low frequency, probably due to some geometrical effect of the shoulders, creating coupling between P and S waves around the excitation point.

The dispersion relations corresponding to the heterogeneous ballast resting on a soft soil are plotted in Fig. 12.13 and correspond to the snapshots of Fig. 12.9. Two important effects can be observed: (i) the propagating behavior is dispersive in the low frequency range, with a very strong slowing of the wave with increasing frequency, similar to that observed in the experimental work of Jacob et al. (2008) on unconsolidated granular packings, and (ii) the dispersion relation vanishes at higher frequencies, which is compatible with the localization effect discussed in the previous section. Note that it does not mean that there is no energy in the system,

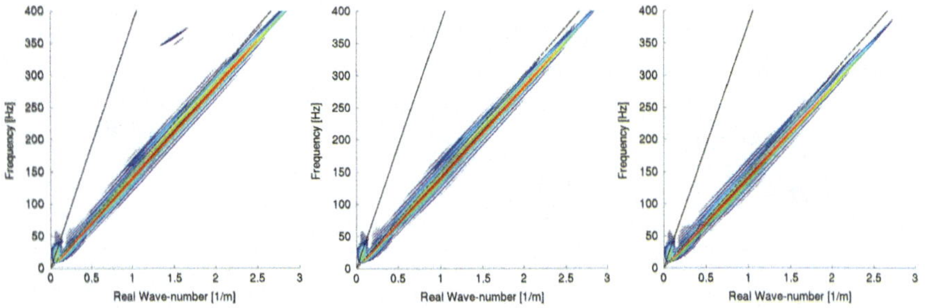

(**a**) Top of ballast. (**b**) Middle ballast interface. (**c**) Soil-Ballast

Fig. 12.12 Dispersion curves for the homogeneous ballast resting on soft soil, in three positions within the section of the ballast. Dispersion relations for the shear wave (*black dotted line*) and pressure wave (*black solid line*) in the ballast and soil are also represented

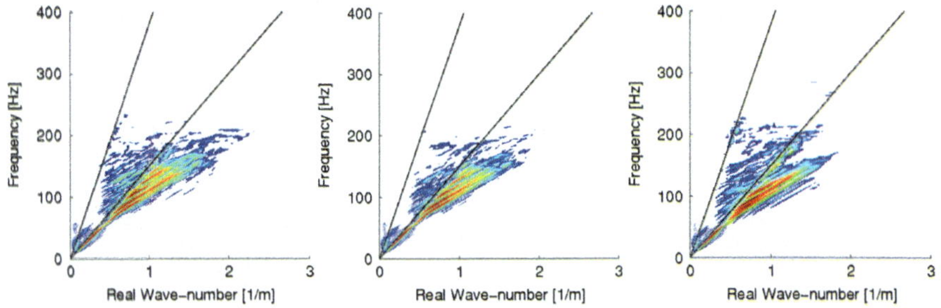

(**a**) Top of ballast. (**b**) Middle ballast interface. (**c**) Soil-Ballast

Fig. 12.13 Dispersion curves for the heterogeneous ballast resting on soft soil, in three positions within the section of the ballast. Dispersion relations for the shear wave (*black dotted line*) and pressure wave (*black solid line*) in the ballast and soil are also represented

but only that the energy does not propagate. We also seem to observe some energy along the P-wave dispersion curve, which is probably due to a strong coupling of P and S wave on the heterogeneities of the ballast.

12.3.4.2 Dispersion Curves for the Ballast Resting on Stiff Soil

Considering now the homogeneous ballast resting on a homogeneous stiff soil, the dispersion relation is plotted on Fig. 12.14, and corresponds to the snapshots of Fig. 12.10. We clearly observe the classical modal behavior of a waveguide (Lagasse 1973). Indeed, a series of thresholds appear, above which new dispersive modes appear, with velocities asymptotically approximating the S wave velocity in the ballast at high frequencies. Note also that, at the interface between the soil and the ballast, both shear wave velocities are retrieved, corresponding to modes either in the ballast or in the soil.

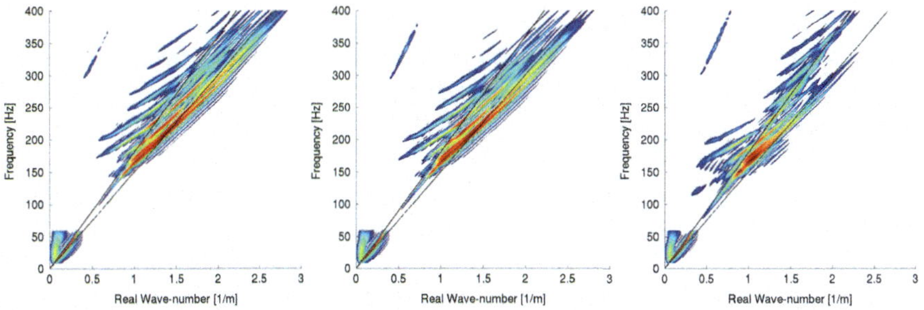

(a) Top of ballast. **(b)** Middle ballast interface. **(c)** Soil-Ballast

Fig. 12.14 Dispersion curves for the homogeneous ballast resting on stiff soil, in three positions within the section of the ballast. Dispersion relations for the ballast shear wave (*black dotted line*) and soil shear wave (*black solid line*) in are also represented

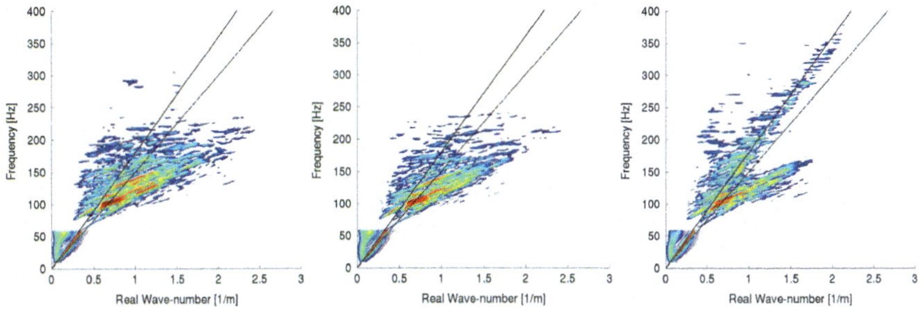

(a) Top of ballast. **(b)** Middle ballast interface. **(c)** Soil-Ballast

Fig. 12.15 Dispersion curves for the heterogeneous ballast resting on stiff soil, in three positions within the section of the ballast. Dispersion relations for the ballast shear wave (*black dotted line*) and soil shear wave (*black solid line*) are also represented

Finally, we plot on Fig. 12.15 the dispersion relation for a heterogeneous ballast resting on a stiff soil. This case corresponds to the snapshots of Fig. 12.11. The behavior is extremely similar to that of the heterogeneous ballast over a soft soil. Hence the complex behavior seems to be mainly controlled by the heterogeneity within the ballast rather then the impedance contrast between the ballast and the soil.

12.4 Conclusions

In this paper, we constructed large scale randomly-fluctuating continuum models of ballasted railway tracks. These models allowed to observe the influence of a passing train as well as to construct dispersion curves for the ballasted layers. The importance of the heterogeneity was stressed very strongly. In particular, it was shown to create relatively slow waves at medium frequencies and even to induce localization at higher frequencies. It was also shown that the heterogeneity of the ballast seems relatively more important that the heterogeneity between ballast and soil. However, this statement should probably be confirmed through a larger set of numerical experiments.

Because of the geometry of the ballast, and because of its heterogeneity, it is absolutely impossible to solve the dynamical equations analytically. Computer simulations are therefore an irreplaceable and unavoidable tool to understand the dynamical behavior of the soil-ballast system. Understanding that behavior, and using more computer simulations, it is expected that improvements can be proposed to mitigate degradations of the ballast and/or invent new maintenance strategies.

References

Aki K (1969) Analysis of the seismic coda of local earthquakes as scattered waves. J Geohysical Res 74(2):615–631

Al Shaer A, Duhamel D, Sab K, Foret G, Schmitt L (2008) Experimental settlement and dynamic behavior of a portion of ballasted railway track under high speed trains. J Sound Vib 316(March):211–233

Alleyne D, Cawley P (1991) A two-dimensional Fourier transform method for the measurement of propagating multimode signals. J Acoust Soc Am 89(3):1159–1168

Arlaud E, D'Aguiar S, Balmes E (2014) Validation of a reduced model of railway track allowing long 3D dynamic calculation of train-track interaction. Comput Methods Recent Adv Geomech 193:1193–1198

Bagi K (2003) Statistical analysis of contact force components in random granular assemblies. Granular Matter 5(1):45–54

Bian X, Cheng C, Jiang J, Chen R, Chen Y (2015) Numerical analysis of soil vibrations due to trains moving at critical speed. Acta Geotech 11(2):281–294

BITRE (2003) Rail Infrastructure Pricing: Principles and Practice. Technical report, Bureau of Infrastructure, Transport and Regional Economics

Cohen G (2001) Higher-Order Numerical Methods for Transient Wave Equations., Scientific Computation Springer, New York

Connolly DP, Kouroussis G, Laghrouche O, Ho CL, Forde MC (2014) Benchmarking railway vibrations - track, vehicle, ground and building effects. Constr Build Mater

Correa L. de Abreu, Cottereau R, Quezada JC, Costa d'Aguiar S, Voivret C (2016b) Randomly-fluctuating heterogeneous continuum model of a granular medium. Comput Mech (submitted for publication)

Drescher A, De Jong GD (1972) Photoelastic verification of a mechanical model for the flow of a granular material. J Mech Phys Solids 20(5):337–340

Festa G, Vilotte JP (2005) The Newmark scheme as velocity-stress time-staggering: an efficient PML implementation for spectral element simulations of elastodynamics. Geophys J Int 161(3): 789–812

Freitas da Cunha, JP (2013) Modelling of Ballasted Railway Tracks for High-Speed Trains. Ph.D. thesis, Universidad do Minho

Goddard JD (1990) Nonlinear elasticity and pressure-dependent wave speeds in granular media. Proc R Soc A Math Phy Eng Sci 430(1878):105–131

Guerin N (1996) Approche expérimentale et numérique du comportement du ballast des voies ferrées (in French)

Guillot L, Aubry L, Le Piver F, Mariotti C, Sèbe O, Thauvin E, Odonbaatar C, Ulziibat M, Demberel S, Sukhbaatar S (2014) Numerical simulation of seismic wave propagation: site effects. Chocs 45:29–36

Hoang TM, Alart P, Dureisseix D, Saussine G (2011) A Domain Decomposition method for granular dynamics using discrete elements and application to railway ballast. Ann Solid Struct Mech 2(2–4):87–98

Jacob X, Aleshin V, Tournat V, Leclaire P, Lauriks W, Gusev VE (2008) Acoustic probing of the jamming transition in an unconsolidated granular medium. Phys Rev Lett 100(158003):1–4

Jehel P, Cottereau R (2015) On damping created by heterogeneous yielding in the numerical analysis of nonlinear RC frame elements. Comput Struc 154:192–203

Jia X, Caroli C, Velicky B (1999) Ultrasound propagation in externally stressed granular media. Phys Rev Lett 82:1863–1866

Knopoff L (1964) A matrix method for elastic wave problems. Bull Seismol Soc Am 54(1):431–438

Komatitsch D (2005) The spectral-element method in seismology. Geophys Monogr Ser 157(55):205–227

Lagasse PE (1973) Higher-order finite-element analysis of topographic guides supporting elastic surface waves. J Acoust Soc Am 53(4):1116–1122

Lawney BP, Luding S (2014) Frequency filtering in disordered granular chains. Acta Mech 225:2385–2407

Leibig M (1994) Model for the propagation of sound in granular materials. Phys Rev E 49(2): 1647–1656

Liu C-H, Nagel SR, Schecter DA, Coopersmith SN, Majumdar S, Narayan O, Witten TA (1995) Force fluctuations in bead packs. Science 269(5223):513–515

Paludo LC, Bouvier V, Cottereau R (2016) Scalable parallel scheme for sampling of Gaussian random fields over large domains. Int J Numer Meth Engrg, In preparation

Rosenblatt M (1952) Remarks on a multivariate transformation. Ann Math Stat, 470–472

Sheng X, Jones CJC, Thompson DJ (2004) A theoretical study on the influence of the track on train-induced ground vibration. J Sound Vib 272:909–936

Shinozuka M, Deodatis G (1991) Simulation of stochastic processes by spectral representation. Appl Mech Rev 44(4):191–204

Suiker ASJ (2002) The Mechanical Behaviour of Ballasted Railroad Tracks. Delft University of Technology, TU Delft

Ta Q-A, Clouteau D, Cottereau R (2010) Modeling of random anisotropic elastic media and impact on wave propagation. Eur J Comput Mech 19(1–3):241–253

Wellmann C, Wriggers P (2012) A two-scale model of granular materials. Comput Methods Appl Mech Eng 205–208:46–58

Chapter 13
Numerical and Parametric Study of MVGs on a UAV Geometry in Subsonic Flow

Miguel Chavez, Silvia Sanvido, Oliver M. F. Browne and Eusebio Valero

Abstract This work details how numerical methods can be employed to determine the optimal arrangement of multiple micro vortex generators (MVG) installed on a UAV wing with its flaps deflected/deployed. For very little cost, the use of MVGs can substantially improve the aerodynamic performance of the UAV by increasing the lift and reducing the drag, particularly during critical phases of the flight (takeoff and landing). The optimal position, related to the detachment zone, and height, related to the boundary layer, of a vortex generator is studied numerically by solving the three dimensional Navier-Stokes equations in a simplified configuration. The numerical study includes a reference baseline configuration. A parametric study is then performed to determine, which configuration gives the maximum gain—highest lift/drag ratio. A preliminary numerical study of convergence for different grids to ensure convergency of the solutions is included. Finally, considering these results, a selection of the optimal configurations based on the aerodynamic performance and the manufacturing limitations.

13.1 Introduction

Nowadays, the control of flow separation is one of the main objectives of the aeronautical industry. This kind of flows appears at the limits of the flight envelope, in particular in low speed and high-lift configurations, and defines the maximum angle of attack and maximum lift. Those are main key design variables to dimension the complicated high-lift devices. Therefore, improving the performance of the wings

M. Chavez (✉)
School of Aeronautics, Universidad Politecnica de Madrid,
Plaza Cardenal Cisneros 3, 28040 Madrid, Spain
e-mail: m.chavez@upm.es

S. Sanvido · O.M.F. Browne · E. Valero
ETSIA-UPM (School of Aeronautics), Universidad Politecnica de Madrid,
28040 Madrid, Spain
e-mail: oliver.browne@upm.es; silvia.sanvido@upm.es

P. Diez et al. (eds.), *Computational Methods and Models for Transport*,
Computational Methods in Applied Sciences 45,
DOI 10.1007/978-3-319-54490-8_13

or flaps would have a positive impact on reducing weight and drag or increasing lift of current passenger aircrafts.

One of the simplest solution currently investigated by the scientific community is the use of flow control devices. These devices aim to modify the aircraft aerodynamic by introducing small modification in the flow, depending on their objective can be used to e.g.: control or provoke separation, reduce buffet, delay the transition to turbulence or for turbulent drag reduction (see Abbas et al. 2013 for a general review). For the particular case of flow detachment, passive flow vortex generators (VG) are becoming a popular alternative to reduce or control the separation on wings or high-lift devices. VG are small protuberances mounted on the wing close to the separation point, they don't require any power input to operate, making the balance energy more positive. Although the detailed explanation of their effect on the surrounded flow physics is not clearly understood, they energize the boundary layer by creating a series of complicated counter rotating vortices, which increase the turbulence intensity and stimulate the entrance of high velocity fresh air from the external flow into the boundary layer.

However, given the large number the parameters involved in the design of these passive devices, such as size, position and orientation with the stream flow and boundary layer, from a practical implementation of these devices, it is necessary to perform a parametric study to find the optimal configuration. This is the topic of this paper.

The effect of different parameters (device height; length; spacing; and stream-wise distance) involved in the design was discussed in Lin (2002). There were different MVG geometries like a wedge, vanes, wishbone, etc. (Fig. 13.1) (Ashill et al. 2005) which were tested experimentally but without a complete parametric study. There is research which has focused on vortex characterization with subsonic flow on a flat plane (Ashill et al. 2002; Yao et al. 2002). Boundary-layer flow-separation control effectiveness has been studied by others researcher (Lin et al. 1990a, b, 1991; Lin 2002; Jenkins et al. 2002) with adverse gradients at low speeds where the MVGs lay on a backward-facing ramp. The results showed the effectiveness of the low-profile VG (completely immerse into the boundary layer) in eliminating the inflection point in the pressure distribution. Taller VG's reduce, also, the detached area but at the cost of causing a strong modification of the input flow, creating 3D flow structures with a subsequent drag penalty. In any case, it is important to emphasize the strong influence and sensitivity of the different parameters involved in their configurations in order to obtain the optimum results.

So far the studies noted above have focused on the interaction MVG/flow with different geometries and different flow speeds. These works allow one to obtain useful knowledge of how to implement these passive devices in different industrial applications. Some of these applications are experimentally tested on airfoil/wing aerodynamic performances enhancement using MVG which increased lift and/or reduced drag for high lift airfoils (Liu et al. 1999; Lin et al. 1994; Klausmeyer et al. 1996; Valarezo 1993), low Reynolds number airfoil (Kerho et al. 1993), highly swept wings (Ashill et al. 1995; Ashill and Fulker 1999), and a transonic airfoil

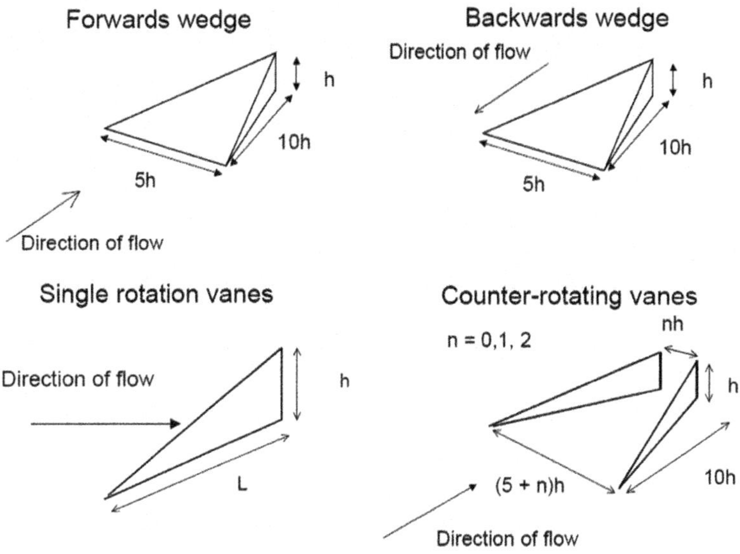

Fig. 13.1 MVGs Geometries, adapted from Ashill et al. (2005)

(Ashill et al. 2001). As for numerical studies, a NACA0012 airfoil (Shan et al. 2008) and an airfoil-flap model (Jansen 2012) have been simulated.

Finally, in addition to the energization of the boundary layer with passive elements in subsonic flows, MVGs in transonic wings and supersonic inlets are capable to alleviate or overcome the adverse effects of separated shock/boundary layer interactions. Some experimental researchers have study shock-induced separation over a flat plate (McCormick 1993) but the major part of the high speed studies have focused on the shock wave control produced inside the engines due to the section change (Lu et al. 2011; Liu et al. 2013; Nolan and Babinsky 2011; Lu et al. 2010a). Additionally, these devices are also employed for noise reduction in the interior aircraft in transonic cruise (Holmes et al. 1987), inlet flow distortion reduction within compact ducts (Anderson et al. 1999; Hamstra et al. 2000), a more over efficient overwing fairing (Tai 2002) and the ground effect over inverted wing in a race car (Kuya et al. 2009).

The research presented in this paper focuses on a UAV geometry with a landing configuration in a subsonic flow, based on Liu (Lin et al. 1994) where the experimental study presents the effect of MVG in lift and drag on a multi-element airfoil. The main objective is to leverage the potential of CFD in terms of simulating different configurations to get the optimal one through a parametric study.

The paper is organized as follows: In Sect. 13.2, a definition of the geometry and flow condition of the problem. A short introduction of the DLR TAU Code used in the simulations in Sect. 13.3. Then in Sect. 13.4 a critical evaluation grid sensitivity has been performed with the baseline model which then allows Sect. 13.5 to focus on the boundary layer thickness and the detached flow area. Section 13.6 describes the cases

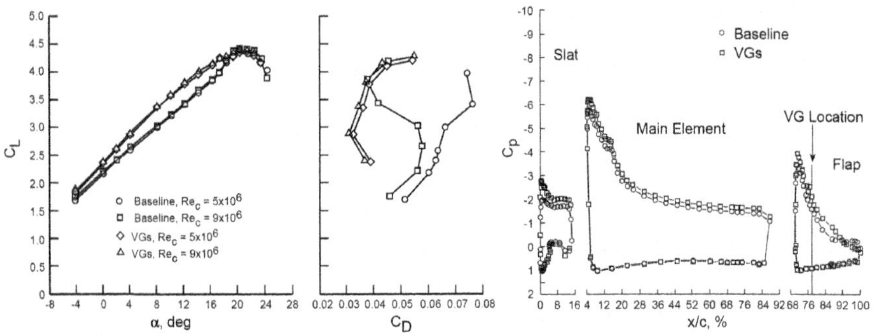

Fig. 13.2 Effect of low-profile VGs on lift, drag and pressure distributions Lin et al. (1994) ($\alpha = 8°$, $M_{inf} = 0.2$, $Re_c = 9 \times 10^6$)

matrix with the different MVGs aspect ratio and subsequently Sects. 13.7 and 13.8 show the optimal configuration and the MVG effects on aerodynamic coefficients, detached flow area and pressure coefficient distribution. Finally, Sect. 13.9 extracts the conclusions from the study (Fig. 13.2).

13.2 Simulation Setup

The objective of this study is the design of a distribution of VG conveniently arranged along the flap to improve the aerodynamic performance in the landing operation of the UAV. The landing is performed with an angle of attack which belongs to the linear part of the lift curve and a flap deflection of 45°. The UAV geometry is composed of a main wing and a flap provided by Unmanned Solutions S.L. (USol). This configuration allows one to study how MVGs affect the boundary layer and obtaining the proper MVG setup where the flow is almost all turbulent.

13.2.1 Flow Condition

The different configurations have been solved with a subsonic turbulent flow and a Reynolds number of 1.52×10^6 (based on a reference surface of 0.168 m) and a velocity of 20.2 m/s. This flow condition belongs to UAV landing configuration which has a slow velocity compared with a commercial airline aircraft. This speed induces to have more laminar flow on the UAV wing upper surface than the commercial airline wing plane. Despite this, the simulation has been resolved as turbulent flow because it is very difficult to accurately produce a fully laminar flow in a wind tunnel.

13.2.2 K2C Model Geometry

The UAV geometry belongs to USol, which has been simplified to generate a valid mesh, and it is a body-wing (Fig. 13.3 left). The wing is composed of a main body, a flap, a spoiler and a wingtip. This configuration has a spoiler fixed to the main body wing while the slotted flap is deflected 45°.

The study of the effect of MVG on aerodynamic coefficients has been done with a single device, which has been installed on the flaps upper surface. Finally has been taken a section of the wing (Fig. 13.3 right) with flap where the study has been done. It allows us to reduce the difficulty of mesh generation and computational time. The reference area for dimensioning the aerodynamic coefficient is $S_{ref} = 0.168$ m.

13.2.3 MVG-Forward Wedge Geometry

The selection of the element is based on the effects of aerodynamic characteristics, the difficulty of meshing the MVG and the manufacture and installation of it over the aircraft. The mesh generation must carefully considered to ensure that the MVG will be immersed inside the boundary layer, and as a result the mesh cells have to be much smaller than the cells at the wing surface. Finally, the manufacture and installation is also considered because once the optimal configuration is identified, the next step will be to test the MVGs over flap of the UAV.

After commenting on the state of the art of MVG, the forward wedge (Fig. 13.4) has been chosen. The parameterization of the MVG is based on a height (h) and position with respect to leading edge of the flap (X). These parameters will be defined by the boundary layer thickness (δ) and the flow detachment position (X) of baseline model flap. Each parameter will be set to 3 different values, so an array of cases with 9 different MVGs will be made, which allows us to obtain the optimal configuration on flap.

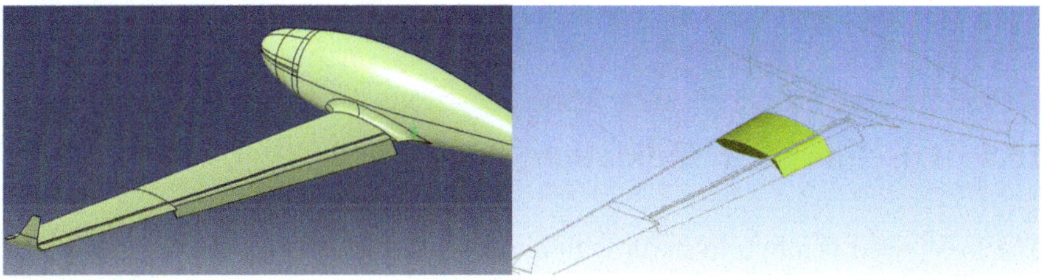

Fig. 13.3 UAV geometry

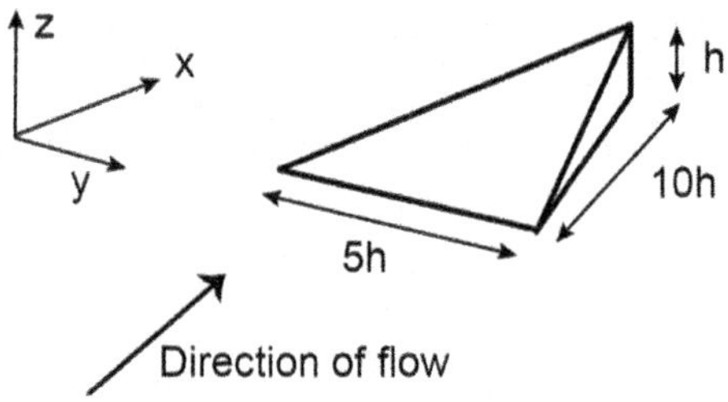

Fig. 13.4 Forward wedge

13.3 Numerical Methodology

The validation has been conducted with the DLR TAU Code, which is used as a standard code within the European aeronautical industry. The DLR TAU Code solves the Reynolds Average Navier-Stokes (RANS) equations on unstructured hybrid grids employing a second order finite volume discretization. Convective terms are discretized with central fluxes and a scalar Jameson-Schmidt-Turkel (JST) numerical dissipation model.

The turbulence models implemented within the TAU code include a linear, as well as non-linear, eddy viscosity models spanning both one and two equation model families. The standard one equation turbulence model in TAU is the Spalart-Allmaras Original (SAO) model, yielding highly satisfactory results for a wide range of applications while being numerically robust. Inside of two equation family exists the $k\omega$ turbulence model has given reasonably good results for free-shear-layer flows with relatively small pressure gradients. It would be an inappropriate choice for problems such as inlets and compressors. Therefore, all the configurations of this study have been simulated SAO because of its desirable properties and stability for external flow.

The solver uses an edge-based dual-cell approach, i.e. a vertex-centered scheme, where inviscid terms are computed employing either a second-order central scheme or a variety of upwind schemes using linear reconstruction (of the left and right states) for second-order accuracy. Viscous terms are computed with a second-order central scheme. Scalar or matrix artificial dissipation may be chosen.

13.4 Baseline Mesh Validation

The mesh used is a hybrid mesh with quad elements for boundary layer zone and triangular elements for the rest of the domain. The structured mesh has 30 elements in the wall normal direction where the first cell near the wall has a height of 2×10^{-5} m providing an $y^+ \approx 1$. This refinement allows us to accurately capture the physics of the boundary layer. Three different grids were used for mesh validation and to study the convergence (Fig. 13.5) at 10°. The coarse grid was created with ANSYS

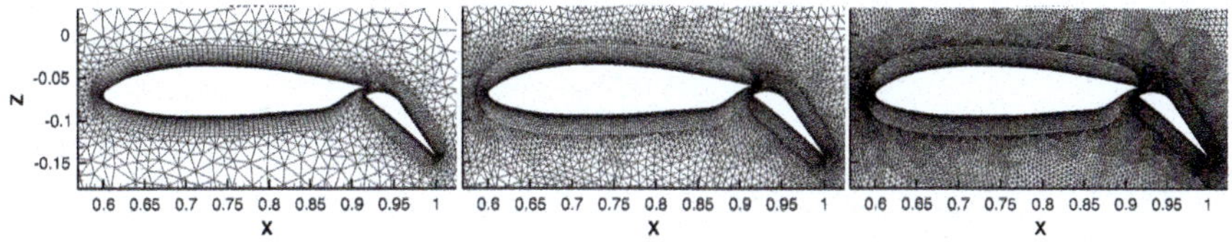

Fig. 13.5 Coarse, medium and fine grid for the grid validation

Table 13.1 Total elements and computational time values for the validation grids on baseline model at 10°

$\alpha = 10°$	Coarse	Medium	Fine
Total elements	5.323.603	27.907.784	164.538.112
Comp. time (h)	9.45	49.5	78.5

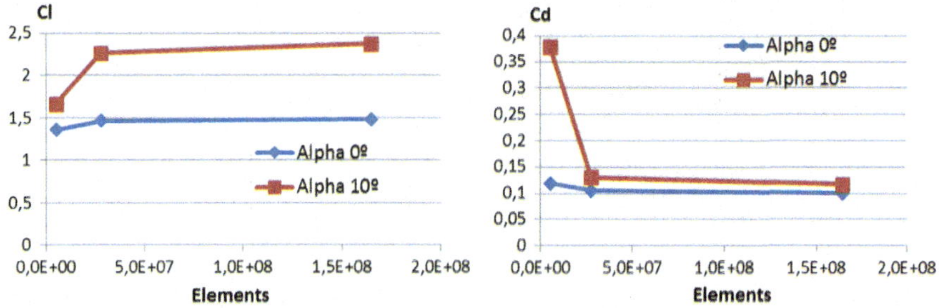

Fig. 13.6 Lift and drag coefficient from Coarse, Medium and Fine grid at 0° and 10°

and the medium and fine grids were generated using the TAU adaptation module which divides the cells in smaller elements. Table 13.1 shows number of elements and the computational time solved at 10°. The computation has been made with MPI calculation in 5 nodes with 12 processors for each angle of attack.

The three grids were used with the Spalart-Allmaras turbulence model. In Fig. 13.6 it can seen that the convergence of the aerodynamic coefficients as a function of the number of elements. In the present study, the simulation of the rest of the configurations has been solved with medium mesh, which guarantees mesh independent results and an acceptable computational effort.

13.5 Baseline Results

Once the grid has been validated, the next step is to solve whole polar curve to predict the point at which stall occurs. Figure 13.7 shows that the stall angle of attack is around 11°. Besides showing aerodynamics coefficients, it is important focus on the behavior of the boundary layer, especially the boundary layer thickness (δ) and

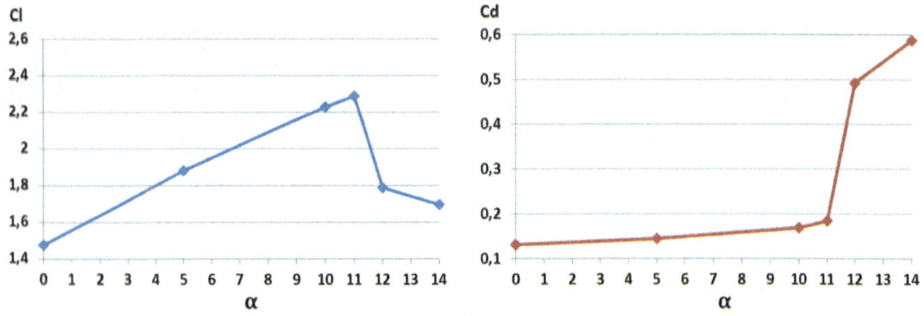

Fig. 13.7 Lift and drag coefficient curve of Baseline model solverd with Medium mesh

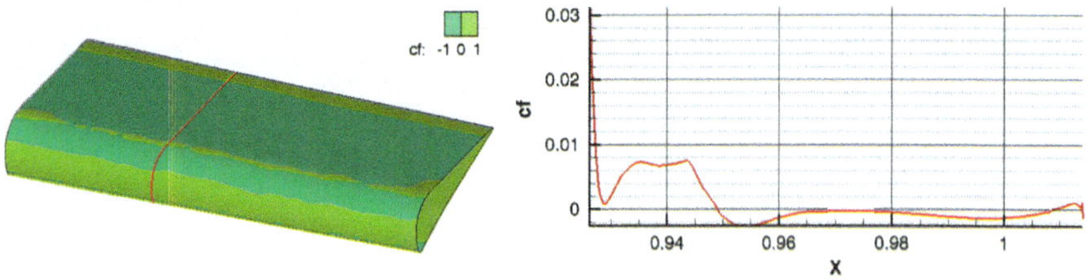

Fig. 13.8 Skin friction coefficient distribution over flap of Baseline model and section $Y = -0.65$ m (*right*) with Medium mesh at $\alpha = 10°$ (*left*)

flow detachment point ($X_{detached}$). The first one defines the h parameter which will change the MVG aspect ratio (Fig. 13.4) and the second one defines the position with respect to the flap leading edge. The MVG position on the flap upper surface will be defined by these two parameters.

The detachment flow point with the distribution of skin friction coefficient over the wing surface has been obtained, where the skin friction coefficient will be negative when the flow is separated. An angle of attack $\alpha = 10°$ has been selected to define this detachment area which is close to the stall and which is also the zone selected to improve the aerodynamics characteristics. A 3D plot is used to display the detachment flow (Fig. 13.8) where the negative (dark green) and positive (light green) skin friction coefficient distribution are drawn over the flap surface. The plot shows the skin friction coefficient value on the flap section ($Y = -0.65$ m) which allows us to obtain, at a glance, the approximate distance between the flow detachment point and the leading edge of the flap, where $X_{dettached} \approx 25\%\ L_{flap}$ ($L_{flap} = 0.08$ m). Also, after the detachment the flow reattaches because there is fluid that comes from the wing upper surface, but the flow again returns to detach in the trailing edge.

The boundary layer thickness has been obtained as a first approximation through a middle section in the flap ($Y = -0.7$ m) where the MVG will be installed. Figure 13.10 represents the x velocity profile just before the flow detaches and is used to obtain a value as a reference and defining parameter h of MVG. The boundary layer thickness is defined until the maximum velocity, in this case 30 m/s. Also, the plot shows the flow behavior where a shear layer develops (Fig. 13.9) due to the fact that the flow comes from the upper wing surface which decelerates the flow. Finally, the flow accelerates again because of the free stream. As can be seen from Fig. 13.10, an approximation value for the boundary layer thickness is $\delta = 1.2$ mm. However, this value is compared with the theoretical expression (M.White 2003) Fluid mechanics. McGraw-Hill for turbulent boundary layers over a flat plate (13.1), which is function of the distance downstream from the start of the boundary layer, $x = 0.032$ m and number Reynolds, $Re_x = \rho \cdot u_0 \cdot x/\mu = 4.61 \times 10^4$. The value of obtained is 1.45 mm, which is twice the value obtained from the x velocity profile. This value is of the same order as the first approximation ($\delta = 1.2$ mm), but it must be remembered that it comes from an expression for a flat plate. Furthermore, in this configuration the shear layer appears which could decrease slightly the boundary layer thickness. For the study this approach is considered acceptable and with consideration to the above mentioned comments, the value was fixed to 1.2 mm to define the different MVGs height (h).

$$\delta \approx \frac{0.382 \cdot x}{Re_x^{\frac{1}{5}}} \tag{13.1}$$

13.6 MVG Geometry Setup

This section shows the 9 different MVG configurations which are a function of the two parameters obtained above. The height (h) defined by the boundary layer thickness ($\delta \approx 1.2$ mm) and position with respect to leading edge of the flap (X) is defined by flow detachment position ($X_{detached} \approx 25\%\ L_{flap} = 20$ mm) over the flap baseline. Table 13.1 shows the three different values for each parameter. Finally, 9 different configurations have been generated to study the effect of MVG, with three values for h and three values for X. h_1 value is dipped into the boundary layer, h_2 value is the approximation value of boundary layer thickness and h_3 value protrudes over the boundary layer. The definition of X follows the same rule, X_1 value is fixed before the flow has detached, X_2 value is close to the detached flow area and X_3 value before the flow detached.

The combination between of all the values permits a study in efficient way to analyse the effect of the MVG aspect ratio and MVG position in the flap over the aerodynamic coefficients (Fig. 13.11 and Table 13.2).

In reference to the mesh generation, 9 different grids for each configuration have been created, where the grid generation methodology slightly has changed to

Table 13.2 Height (h) and position with respect to LE (X)

Height (mm)		Position (mm)	
h_1	0.375δ	X_1	$0.48X_{det}$
h_2	δ	X_2	$0.9X_{det}$
h_3	1.5δ	X_3	$1.24X_{det}$

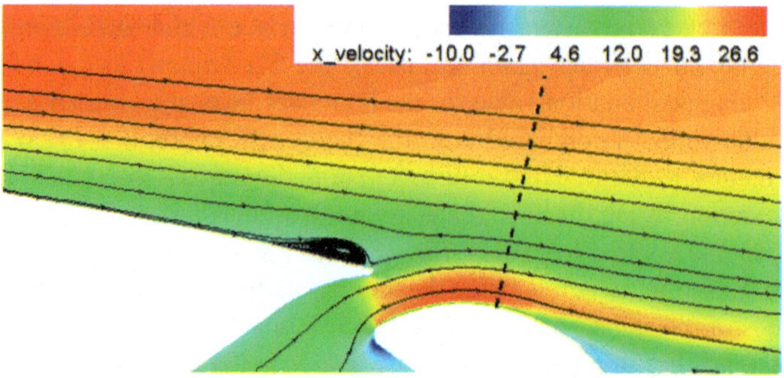

Fig. 13.9 X velocity field (Section $Y = -0.7$) and streamlines in the slot flap at 10°

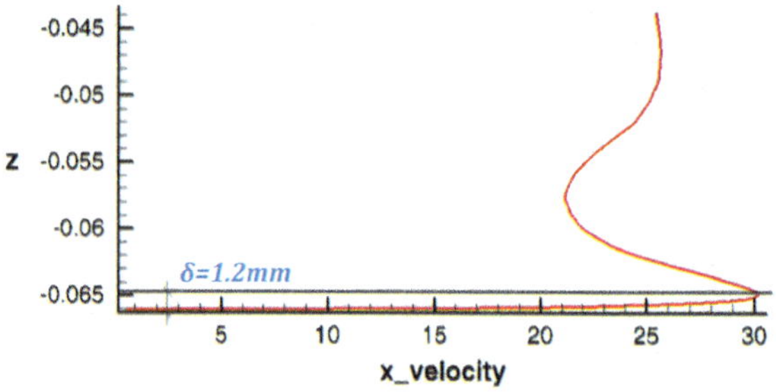

Fig. 13.10 X velocity profile from *dashed line* in Fig. 13.9 at Section $Y = -0.7$

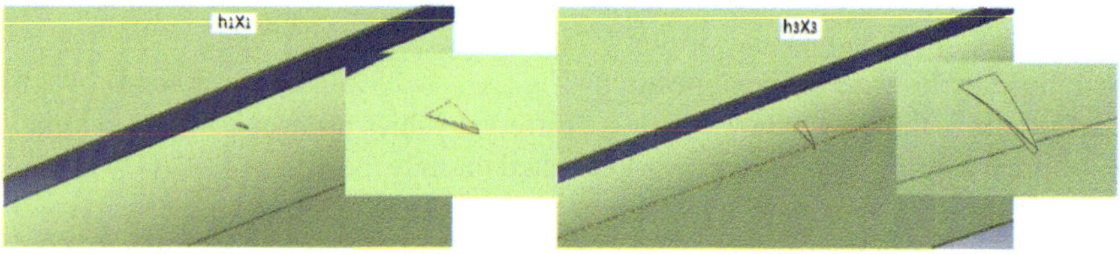

Fig. 13.11 MVG geometry over the flap of h1X1 and h3X3 configuration

reproduce correctly the MVG geometry. The reference length ($c = 0.4$ m) 400 times more than the MVG $1 \times 10e^{-3}$ m, so the elements close to the MVG must be very small to capture global effects produced by the MVG. Finally, a rectangular refined region has been created as shown in Fig. 13.12.

13.7 MVG Results

After finishing the problem definition and mesh validation, the simulation of all cases have been computed using the medium mesh and SAO turbulence model.

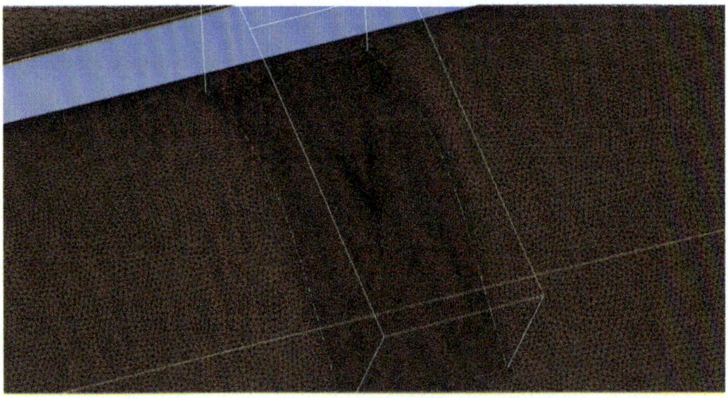

Fig. 13.12 Rectangular refinement grid for h2X2 MVG configuration

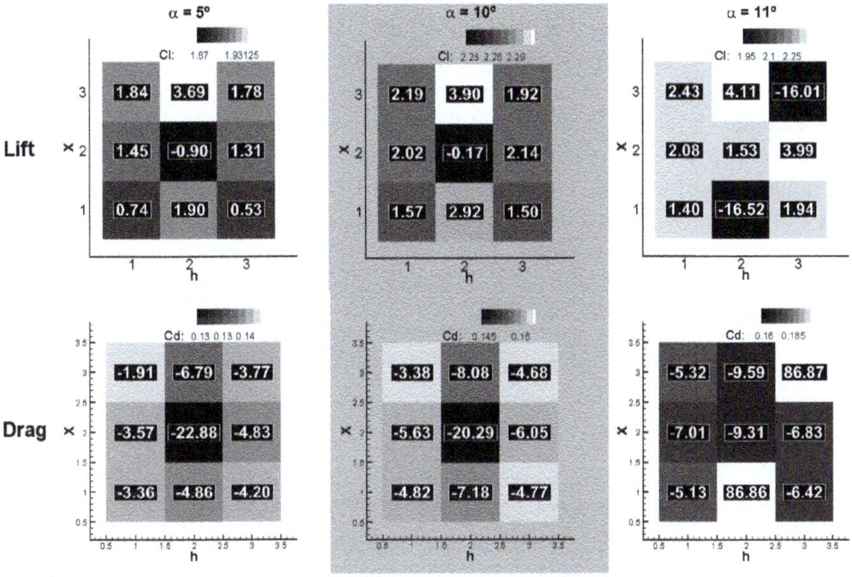

Fig. 13.13 Numerical results of lift (*top*) and drag (*bottom*) coefficient with the increase percentage over the baseline result (number) at 5°, 10° and 11°

Figure 13.13 shows the numerical result in field colors of lift (top raw) and drag (bottom raw) coefficient as a function of the MVG height (h), position with respect to leading edge of the flap (X) and angle of attack (each column, 5°, 10° and 11°, respectively). It also shows the percentage difference in each configuration compared with the baseline result which allows us to determine, at a glance, the optimal configuration and extract reasonable conclusions.

The configurations with h1 and h3 have a similar behavior; an increase in C_L and a decrease of C_D, this effect increases as the X. With h1 this behavior occurs in the over the whole polar, while with h3 at $\alpha C_{Lmax} = 11°$ the model h3X3 goes into stall. This stall is due to the fact that the configuration h3X3, which protrudes over the boundary layer and it is fixed after the detached zone, decrease the energy in the boundary layer producing the flow detaches. The h2 configuration has completely different behavior compared with h1 and h3. Before the detached zone (h2X1) the C_L is higher and C_D is lower than h1X1 and h2X3 configurations, this occurs until the $\alpha C_{Lmax} = 11°$ where model h2X1 goes into stall again. In the detached zone the behavior (h2X2) changes where the C_L decreases slightly and the C_D decreases considerably. Both configurations have consistent behavior in the linear zone, while in the αC_{Lmax} with X1 is at stall and while X2 shows desirable behavior. Finally the configuration h2X3 is even better than h3X2, which has the best aerodynamic qualities in reference with the baseline model. The configuration h2X3 achieves at αC_{Lmax} a C_L increase of 4% and a C_D decrease of 8%, which improve considerably the aerodynamic characteristics. However, the configuration h3X2 couldn't be installed because the height h3 is higher than the minimum gap between wing and flap.

In summary, only the MVG implementation improves the aerodynamic coefficients, increasing the lift because the element generates micro vortex which energizes the boundary layer and it allows the flow to remain attached for longer. This produces a reduction in drag. However, there is some specific configuration that causes a premature stall which is important to avoid. The optimal MVG position over the flap (X) should be after the detachment flow in the baseline model because it is more effective to energize the boundary layer, but the MVG height (h) should be less than the boundary layer thickness because it produces stall, so the optimal is the height of the boundary layer.

13.8 MVG Effects

The optimal configuration is h2X3 which improves the aerodynamic coefficients (Fig. 13.14) and delays the detachment zone (Figs. 13.15 and 13.16) as it was mentioned in the introduction. This section focuses on the optimal configuration to better

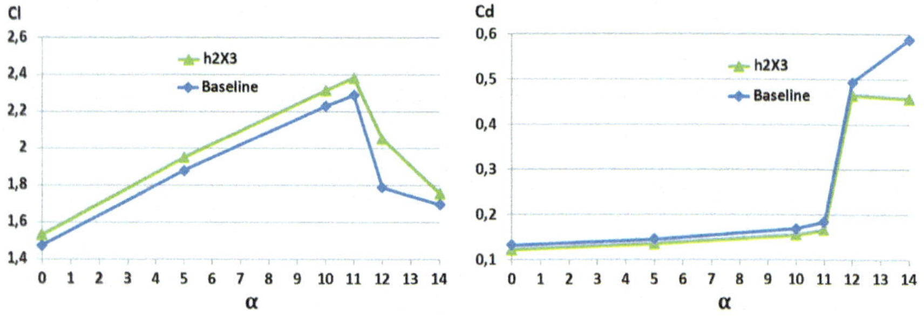

Fig. 13.14 Cl and Cd comparison between baseline and h2X3 configuration

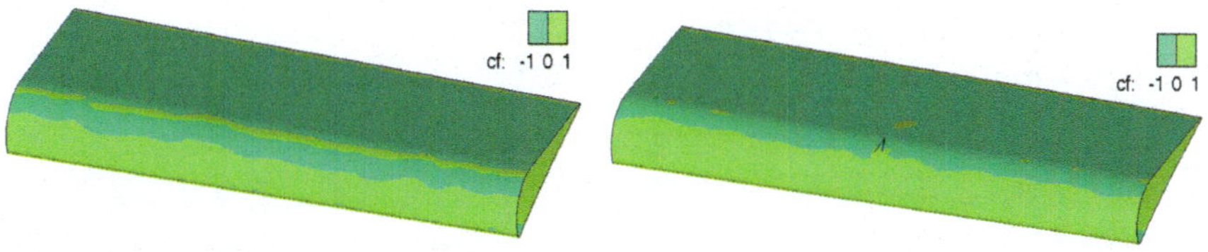

Fig. 13.15 Cf flap surface comparison between baseline model (*left*) and h2X3 configuration (*right*) at = 10°

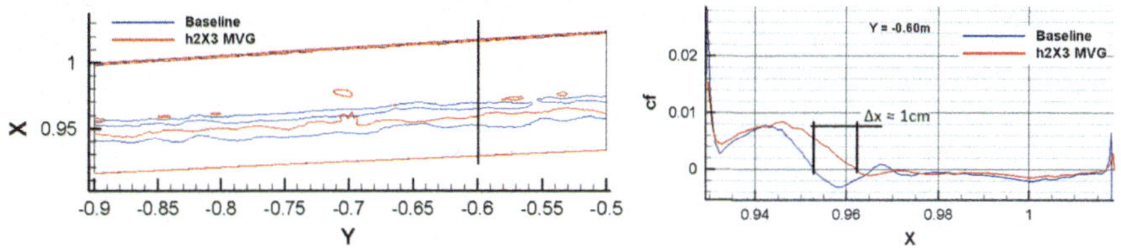

Fig. 13.16 Cf comparison at section $Y = -0.6$ m

understand how the installed MVG affects in the boundary layer. This has been evaluated with the skin friction coefficient and the pressure coefficient. The comparison between the baseline model and the h2X3 configuration with an angle of attach of 10° has been made, which belongs to a landing set up. Figure 13.14 shows the aerodynamic coefficients improvement between the baseline model and the h2X3 MVG configuration. The C_L increases by 4%, including the C_{Lmax}, and the C_D decreases by approximately 8% over the baseline model. The process of stall is smoother which is a more safe way for the UAV.

The skin friction allows the prediction of the detachment flow point on the flap upper surface, where the negative value (dark green) means that the flow is detached. Figure 13.15 shows a skin friction coefficient 3D plot of the baseline flap (left) and h2X3 configuration flap (right) where the MVG implementation gets to delay the detached area and little reattachment area disappears due to the fact that the element energize the boundary layer. However, this reattachment area starts to appear in a little regions far of the MVG which define the effect scope of MVG over the flap. This scope could define the optimal distance between MVGs.

The plot below (Fig. 13.16) has superimposed the detached line of both configurations (left), the baseline flap (blue line) and h2X3 configuration flap (red line). In the right flap section ($Y = -0.60$ m) which allows us to obtain, at a glance, the approximate distance between both detached line. This detached area delays around $0.5\%(X_{detached})_{baseline}$ as can be seen in the left plot. To understand where the aerodynamic coefficients improvement comes, the (Fig. 13.17) shows a Cp distribution comparison of three different sections along the whole wing between baseline model (blue line) and h2X3 configuration (red line). There is relatively little difference on the main wing part due to the MVG effect does not effect the upstream. In the detached area of the flap upper surface, there is an increment of area in the three

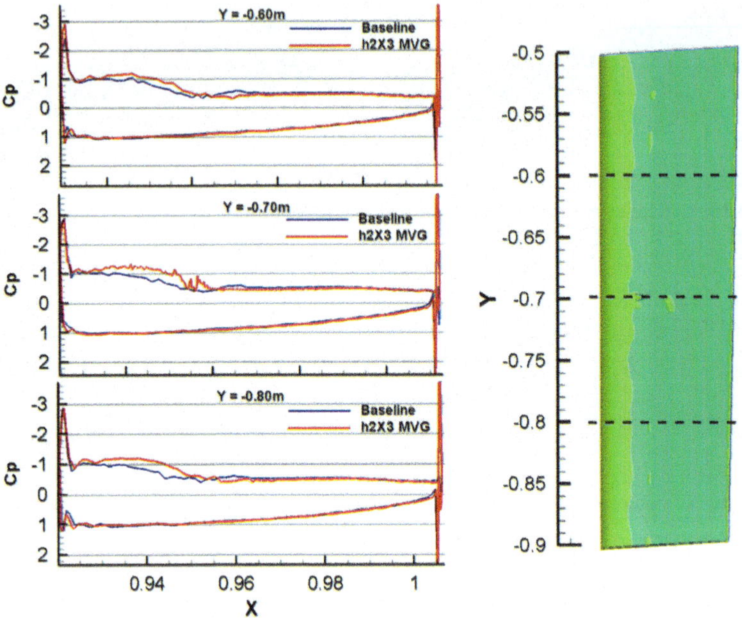

Fig. 13.17 Cp comparison at section $Y = -0.6$ m

sections which directly has an effect on the lift. This effect becomes less significant with increasing distance from the MVG as has been shown in sections. The horizontal zone in the Cp distribution after the MVG effect means that the flows is detached.

13.9 Conclusions

This paper has demonstrated the improvement in aerodynamics coefficients provided by the implementation of a MVG on the flap upper surface. The introduction of a MVG on a slice of the wing in this configuration causes an approximately 4% increase in C_L and a decrease of 7% in C_D comparing it with the lift and drag curve of base line case, also causing a softer stall by comparison, by making it safer. These results are due to the fact that the passive elements keep the flow attached for longer than in the baseline case. This case is able to delay the flow by approximately 0.5% with to respect the $(X_{detached})_{Baseline}$.

The MVGs generate vortexes that transfer low energy fluid from the surface into the mainstream, and bring higher energy fluid from the mainstream down to the surface where the higher kinetic energy level is able to withstand a greater pressure rise before separation occurs. Therefore, the optimal MVG position over the flap (X) should be after the detachment flow in the baseline model because it is more effective to energize the boundary layer, but the MVG height (h) should be less than the boundary layer thickness because it produces stall, so the optimal is the height of the boundary layer.

MVG flap placement improves the behavior on landing or descent without causing any negative effect on other flight stages. The increase in C_L allows decrease lowering speed, causing the decrease in length of landing track to the aircraft. The consumption will also decrease because the flight speed and the drag is reduced. This type of numerical study with passive elements on a UAV in much less expensive as opposed to a flight or wind tunnel test.

Acknowledgements This work was supported by European Comission through the NNATAC GA 324298 project. And the computer resources, technical expertise and assistance provided by the Supercomputing and Visualization Center of Madrid.

References

Abbas A, De Vicente J, Valero E (2013) Aerodynamic technologies to improve aircraft performance. Aerosp Sci Technol 28(1):100–132

Anderson BH, Miller DN, Yagle PJ, Truax PP (1999) A study of MEMS flow control for the management of engine face distortion in compact inlet systems. In: FEDSM99-6920, 3rd ASME/JSME joint fluids engineering conference, San Francisco, CA

Ashill PR, Fulker JL (1999) A review of flow control research at DERA. In: IUTAM symposium on mechanics of passive and active flow control. Springer, pp 43–56

Ashill PR, Riddle GL, Stanley MJ (1995) Separation control on highly-swept wings with fixed or variable camber. Aeronaut J 99(988):317–327

Ashill PR, Fulker JL, Hackett KC (2001) Research at DERA on sub boundary layer vortex generators (SBVGs). AIAA paper, vol 887

Ashill PR, Fulker JL, Hackett KC (2002) Studies of flows induced by sub boundary layer vortex generators (sbvgs). AIAA paper 968:2002

Ashill PR, Fulker JL, Hackett KC (2005) A review of recent developments in flow control. Aeronaut J 109(1095):205–232

Hamstra JW, Miller DN, Truax PP, Anderson BA, Wendt BJ (2000) Active inlet flow control technology demonstration. Aeronaut J 104(1040):473–479

Holmes AE, Hickey PK, Murphy WR, Hilton DA (1987) The application of sub-boundary layer vortex generators to reduce canopy "Mach rumble" interior noise on the Gulfstream III. Rep./AIAA

Jansen DP (2012) Passive flow separation control on an airfoil-flap model

Jenkins L, Gorton S, Althoff, Anders S (2002) Flow control device evaluation for an internal flow with an adverse pressure gradient. AIAA paper 266

Kerho M, Hutcherson S, Blackwelder RF, Liebeck RH (1993) Vortex generators used to control laminar separation bubbles. J Aircraft 30(3):315–319

Klausmeyer SM, Papadakis M, Lin JC (1996) A flow physics study of vortex generators on a multi-element airfoil. AIAA paper

Kuya Y, Takeda K, Zhang X, Beeton S, Pandaleon T (2009) Flow separation control on a race car wing with vortex generators in ground effect. J Fluids Eng 131(12):121102

Lin JC, Howard FG, Bushnell DM, Selby GV (1990a) Investigation of several passive and active methods for turbulent flow separation control. In: 21st plasma dynamics and lasers conference AIAA, fluid dynamics, vol 1

Lin JC, Howard FG, Selby GV (1990b) Small submerged vortex generators for turbulent flow separation control. J Spacecr Rockets 27(5):503–507

Lin JC, Selby GV, Howard FG (1991) Exploratory study of vortex-generating devices for turbulent flow separation control. AIAA paper
Lin JC (2002) Review of research on low-profile vortex generators to control boundary-layer separation. Progr Aerosp Sci 38(4):389–420
Lin JC, Robinson SK, McGhee RJ, Valarezo WO (1994) Separation control on high-lift airfoils via micro-vortex generators. J Aircraft 31(6):1317–1323
Liu C, Sun Z, Wang X, Yan Y (1999) Control of turbulent boundary-layer separation using micro-vortex generators. In: 30th fluid dynamics conference, american institute of aeronautics and astronautics, vol 1
Liu C, Sun Z, Wang X, Yan Y (2013) The vortical structures in the rear separation and wake produced by a supersonic micro-ramp. In: 51st AIAA aerospace sciences meeting including the new horizons forum and aerospace exposition, vol 1
Lu F, Pierce AJ, Shih Y (2010a) Experimental study of near wake of micro vortex generators in supersonic flow. AIAA paper 4623:2010
Lu FK, Li Q, Shih Y, Pierce AJ, Liu C (2011) Review of micro vortex generators in high-speed flow. In: 49th AIAA aerospace sciences meeting
McCormick DC (1993) Shock/boundary-layer interaction control with vortex generators and passive cavity. AIAA J 31(1):91–96
M.White F (2003) Fluid mechanics. McGraw-Hill
Nolan WR, Babinsky H (2011) Characterization of micro-vortex generators in supersonic flows. AIAA paper 71:2011
Shan H, Jiang L, Liu C, Love M, Maines B (2008) Numerical study of passive and active flow separation control over a NACA0012 airfoil. Comput Fluids 37(8):975–992
Tai T (2002) Effect of micro-vortex generators on V-22 aircraft forward-flight aerodynamics
Valarezo WO (1993) Topics in high-lift aerodynamics. In: 1993 AIAA 24th fluid dynamics conference
Yao C-S, Lin JC, Allan BG (2002) Flow-field measurement of device-induced embedded streamwise vortex on a flat plate. NASA STI/Recon Technical report N 3:12931

Chapter 14
Investigating Side-Wind Stability of High Speed Trains Using High Resolution Large Eddy Simulations and Hybrid Models

Moritz M. Fragner and Ralf Deiterding

Abstract Crosswind flow over high speed trains can pose serious safety concerns for rail transport. Methodologies for evaluating the aerodynamic forces exerted on the train include full-scale measurements, physical modeling using wind-tunnel experiments and numerical modeling using computational fluid mechanics (CFD). Although CFD presents the most cost-effective approach, it faces severe uncertainties in the predicted forces, most of which are related to the turbulence modeling technique employed. In here we investigate the influence of various turbulence modeling approaches on crosswind flow simulations and calculated force coefficients. In particular, we perform URANS, LES and DDES simulations utilizing the DLR Next Generation Train 2 model geometry. Particular emphasis is laid on simulating a wind angle of 30 degrees and Reynolds number of 225,000 for which validation data is provided by wind tunnel measurements. We confirm that a major vortex system on the leeward side of the train develops, which mainly drives the overturning force and moment of the train. The lift force is determined mainly by the underbody flow, which is characterized by unsteady vortex shedding. Due to its dual ability to properly model the roof boundary layer on the one hand and to resolve small-scale turbulent eddies in the underfloor region on the other, the DDES approach is found to give the most accurate force predictions. LES overpredicts the overturning force and moment, while URANS overpredicts the lift force.

M.M. Fragner (✉)
German Aerospace Center, Bunsenstrasse 10, 37073 Göttingen, Germany
e-mail: moritz.fragner@b-tu.de

M.M. Fragner
Numerische Strömungs- und Gasdynamik BTU Cottbus - Senftenberg, Siemens-Halske-Ring 14, 03046 Cottbus, Germany

R. Deiterding
University of Southampton, Highfield Campus, Southampton SO17 1BJ, UK
e-mail: R.Deiterding@soton.ac.uk

P. Diez et al. (eds.), *Computational Methods and Models for Transport*, Computational Methods in Applied Sciences 45,
DOI 10.1007/978-3-319-54490-8_14

14.1 Introduction

Modern high speed trains are of considerable risk of overturning or derailment when operating in the presence of strong crosswinds. In order to assess the safety of such vehicles a detailed understanding of the aerodynamic forces is required. Currently, there are three different methodologies for evaluating the aerodynamics of trains: full-scale measurements, physical modeling using wind-tunnel experiments and numerical modeling using computational fluid mechanics (CFD). Latter has the potential advantage of providing a cost-effective tool to gain a deep understanding of the aerodynamic flow around the vehicle for arbitrary operating and geometry conditions. However, there are still many difficulties associated with CFD, most of which are related to modeling and resolving turbulent dynamics on multiple scales.

In the case of high speed trains operating in crosswinds at large yaw angle, substantial unsteady vortex shedding on the roof and underfloor side of the train is expected. The complexity is even enhanced by the influence of the train nose, which renders the flow truly three-dimensional. Small-scale unsteady effects are furthermore expected to result from the interaction of the flow with the bogie cavity and the wheels. In such a scenario, a simple Reynolds averaged Navier-Stokes (RANS) assumption representing all turbulence effects on the mean flow field with a simple, isotropic model is generally violated and forces derived from this flow field tend to show sizeable discrepancies to experimental values.

In order to better understand the influence of various turbulence modeling approaches on crosswind flow simulations, we have conducted a case study ranging from unsteady RANS (URANS) and delayed detached eddy simulations (DDES) to large eddy simulations (LES). The geometry used in this investigation corresponds to a front car of the DLR's Next Generation Train 2 (NGT2) (Bürkle and Winter 2013), an in-house high-speed concept train specifically designed to foster train-related research, see Fig. 14.1. We focus on simulating yaw angles of $\beta = 30°$, for which the vortex system on the leeward side of the train leads to sizeable uncertainties in predicted integral coefficients. A Reynolds number of 225,000 (based on the height of the train) is used for which experimental wind tunnel measurements of force coefficients are available. The simulations have been conducted using the free open-source software OpenFOAM. Given the low Mach number of typical wind tunnel experiments, we only considered incompressible methods, based on pressure-correction-type algorithms, in this work.

The topic of trains operating in strong crosswinds has been addressed by several previous publications. Experimental studies on the side-wind problem have been undertaken, for example, by Baker (2003), Baker et al. (2004) and by Haff et al. (2012). Numerical studies using RANS based models have been conducted by Diedrichs (2003) and Khier et al. (2000).

Owing to the complexity of the flow, LES simulations have been performed by Hemida and Krajnović (2008, 2010). Here they study the influence of varying yaw angle and nose shape on flow structures using a simplified train model without inter-cap gaps and wheel bogies at a Reynolds number of 300,000. Similar to the flow over a cylinder, they found that vortex shedding is induced by the shear layer instability at the periphery of the re-circulation region. However the nose shape can have a significant influence on both the time-averaged and instantaneous flow, introducing three-dimensional effects to a region of a length of 3.5 train heights from the tip of the nose in the direction of the length of the train. The extend of this region is even increased for elongated nose shapes. Their studies also suggest that there are qualitative differences between low ($\leq 35°$) and larger yaw angles. While for large yaw angles the flow is predominantly characterized by unsteady vortex shedding, the side wind flow at low yaw angles changes from that associated with a steady slender body flow, where pairs of steady line vortices form the wake flow, to unsteady vortex shedding after a distance of about five train heights from the tip of the nose in the direction of the train length.

Dynamic flow patterns are also in agreement with the experiments by Chiu and Squire (1992) that clearly exhibited unsteady vortex shedding at yaw angles of 40°.

In their article, Hemida and Krajnović (2009) confirm this observation using LES simulations with a simplified ICE2 train geometry. They study the flow at a 30° yaw angle and at a Reynolds number of 200,000. Additionally a separate flow regime in the lower part of the wake flow was identified, in which unsteady wake vortices attach and detach from the surface of the train in a regular fashion.

In this context, the experiments by Copley (1987) give further valuable insights. They study crosswind flow at yaw angles of 20°–35° and Reynolds number of $3.7 \cdot 10^5$ and identify vortices forming on the leeward side as a major contributor to the overall aerodynamic forces on the vehicle. When the train was modelled without wheels, a three-dimensional Karman vortex street developed, with vortices being shed from successive points on both the roof and underside of the train. However with wheels included, the vortices shed from the underside of the train were disrupted and reduced in strength, increasing the overall forces on the train.

Crosswind studies around a freight train model were performed by means of LES by Hemida and Baker (2010) at Reynolds number of 300,000. Due to the box-like shape of their container wagon model, they find that the flow already separates at the sharp windward edge to form a large separated flow region on the roof of the container. In addition their results suggest that high speed flow through the gaps between the containers can result in a much more complex flow topology in the wake of the wagon.

The issue of flow through the gaps between containers has also been addressed by Östh and Krajnović (2014), though not within the side-wind context. In their study it was shown that the flow inside the gaps may form counter-rotating vortices, which make the train appear as a single body to the oncoming flow and shields the wagon from any direct impingement of the flow.

The study by Hemida and Baker (2010) furthermore investigates the effect of ground motion on the aerodynamic coefficients. Their findings suggest, that the lift force may be increased substantially when ground motion is considered, while the side force is only marginally affected. This was also investigated by Krajnović et al. (2012) using LES simulations. Although their simulations were performed at rather low Reynolds number (Re = 22,615) their results suggest that aerodynamic forces could exhibit significantly larger values than compared to the stationary model, which may show the limited validity of using stationary vehicles in wind tunnel experiments or CFD simulations.

Most of the turbulence modelling approaches in previous publications are based either on RANS or simply Smagorinsky LES-type models. A whole set of turbulence modeling techniques has been presented by Krajnović (2014) ranging from LES, hybrid RANS-LES to unsteady RANS simulations employing generic train-like vehicles. In their study hybrid RANS-LES simulations have been performed in the form of Partially-Averaged Navier Stokes (PANS) equations. Morden et al. (2015) compared a number of RANS and two DES models against wind-tunnel data for zero degree yaw angle by means of surface pressures. Their results suggest that the DDES approach delivered the highest accuracy compared to the experimental data.

In this paper we will present simulations employing several turbulence modelling approaches for simulation of the aerodynamic flow of trains in strong crosswinds. URANS and DDES simulations based on the Spalart-Allmaras models as well as LES using a simple Smagorinsky model will be presented. The paper is organized as follows. In Sect. 14.2, we will discuss the numerical setup and the details of the turbulence modelling approaches. In Sect. 14.3 the results will be discussed in terms of computed force coefficients, total pressure contours and vorticity slides. The conclusions will be given in Sect. 14.4.

14.2 Setup

The model geometry we employ is the DLR's Next Generation Train 2 (NGT2) (Bürkle and Winter 2013) geometry. This is an in-house high-speed concept train specifically designed for train-related research. It includes bogie and wheels details to mimic a realistic train. We consider the model at 1:25 scale, hence it has dimensions $[L \times H \times W] = [1400 \times 186 \times 122]$ mm. The geometry consists of the front car (for which forces are measured) and a streamlined body intended to replicate the aerodynamic influence of following cars on the lead car, see Fig. 14.1. The gap between the two cars is very narrow at only 1 mm in width. Because the resolution requirements are rather substantial in order to properly resolve this gap region, we have neglected the gap in the numerical simulations. Validation data from the DLR Side Wind Test Facility Göttingen (SWG) is available for the considered Reynolds number (Re =225,000) and yaw angle ($\beta = 30°$). Due to a requirement for the wind-tunnel experiments, the model is mounted onto a splitter plate to reduce the influence of the wind tunnel floor boundary layer.

Forces and moments are calculated for the lead car with respect to the body aligned coordinate system depicted in Fig. 14.2. Force coefficients have been normalized using the dynamic pressure $q = \frac{1}{2}U_\infty^2\rho$ multiplied by the scaled cross sectional area A, where U_∞ is the far field/inlet velocity and the density is $\rho = 1$ (as we consider incompressible simulations). The reference train cross sectional area is

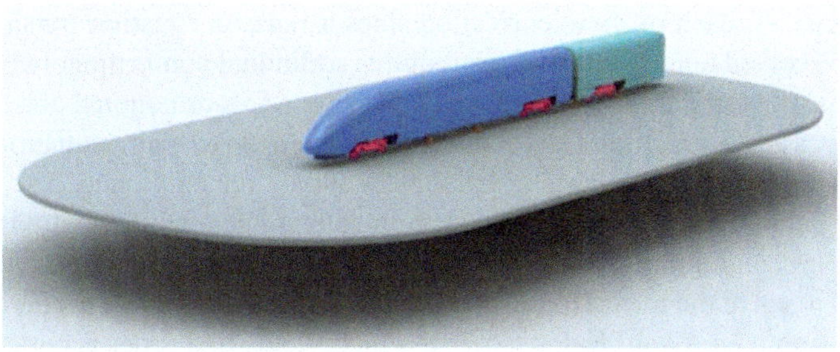

Fig. 14.1 NGT2 front car model with streamlined end body attached to a stationary flat plate as used in the experiment

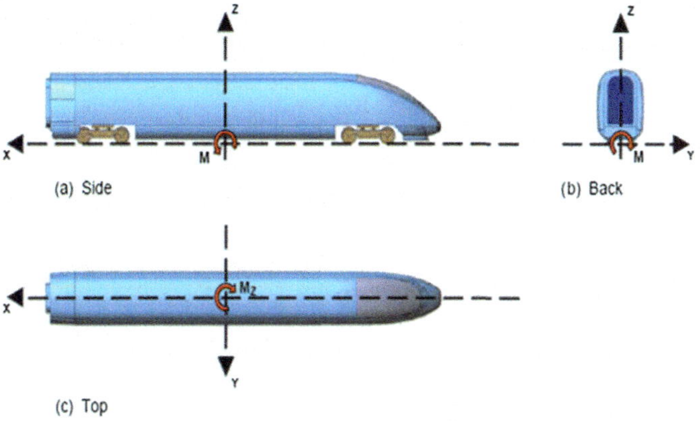

Fig. 14.2 Coordinate system used for force and moment calculations

$A = 0.016\,\text{m}^2$, equivalent to a cross sectional area of $10\,\text{m}^2$ in the 1:1 scale. Moments are defined about the coordinate system origin and are normalized by qAh, with $h = 0.12\,\text{m}$ being the characteristic length scale used in this paper, this being close to the train height.

The geometry is rotated around the z-axis by $\beta = 30°$ and placed in the center of the numerical domain, which is a simple box as depicted in Fig. 14.4. The box has dimensions $[L \times H \times W] = [83h \times 33h \times 42h]$, hence domain walls are sufficiently far away from the train model.

The computational investigations of the model vehicle aerodynamics are being performed with the freely available OpenFOAM software system (Weller et al. 1998; Jasak 1996). Incompressible simulations are conducted using the implicit transient solver *pimpleFoam*, which—like most of the OpenFOAM solvers—is based on pressure corrections algorithms (cf. Ferziger and Peric 2012) to approximate the incompressible Navier-Stokes equations. Because we choose a time step, which yields a maximum Courant number of Co ~ 1 everywhere in the domain, we employ the Piso algorithm (i.e. Pimple algorithm with only one outer iteration) with two pressure correction steps. At each of these correction steps a Poisson equation for the pressure needs to be solved numerically, which requires additional non-orthogonal correctors steps on unstructured meshes. Here we employ four non-orthogonal corrector steps for each pressure correction. Hence, at each time step the pressure equation is solved eight times, while the momentum equation is solved for only once. The set of linearized equations is then solved using a conjugate gradient method with incomplete Cholesky preconditioning for the pressure, while a simple Gauss Seidel solver is employed to solve the linearized momentum equation. The temporal operator is discretized using a backward Euler scheme, which yields first order accuracy in time. The divergence operator uses a linear-upwind scheme, while the Laplacian operator is corrected for mesh non-orthogonality as mentioned above. Both of these operators are discretized to second order in space. Turbulence modelling is considered via the subgrid scale stresses, which are obtained after applying a spacial filter to the incompressible Navier-Stokes equations:

$$\partial_t \bar{\mathbf{u}} + \bar{\mathbf{u}} \cdot \nabla \bar{\mathbf{u}} = -\frac{1}{\rho} \nabla \bar{p} + \nu \nabla^2 \bar{\mathbf{u}} + \nabla \tau^{\mathrm{SGS}} \tag{14.1}$$

$$\nabla \bar{\mathbf{u}} = 0 \tag{14.2}$$

where $\bar{\mathbf{u}}$, $\bar{p}$ are resolved filtered quantities and

$$\tau^{\mathrm{SGS}} = \overline{\mathbf{u}\mathbf{u}} - \bar{\mathbf{u}}\bar{\mathbf{u}} = -2\nu_t \bar{S}, \tag{14.3}$$

where $\bar{S}$ is the symmetric part of the velocity gradient tensor and ν_t is the turbulent viscosity. The appearance of τ^{SGS} reflects the fact that the non-linear and filtering operators do not commute. It is precisely this term which represents the effect of non-resolved scales onto the resolved scales and which requires modelling. It should be noted that although RANS and LES are conceptually different approaches, Eq. (14.1) is formally identical to the unsteady RANS momentum equation, the difference being that all turbulent dynamics are hidden in the stress tensor τ^{SGS} in URANS, while in LES the largest part of turbulent motion is resolved and τ^{SGS} carries only the remaining portion.

For the URANS simulations we employ the standard Spalart-Allmaras (SA) model to describe ν_t in Eq. (14.3). For LES we use a simple Smagorinsky model with van Driest damping function to ensure modelled turbulence disappears next to the wall facing cells. Because of the large resolution requirements in order to properly resolve wall boundary layers with LES at industry typical Reynolds numbers, hybrid models such as detached eddy simulations (DES/DDES) have become a popular alternative to LES (Spalart et al. 1997, 2006). DDES methods employ RANS models for τ^{SGS} near the wall and transition into LES mode further away by manipulating the model length scale of these models. We use the SA model as the underlying RANS model, while the model length scale d entering the SA model, is being modified according to:

Table 14.1 Specifications of the mesh used in the simulations

Properties	
Nr. cells	34 M
Δx	3 mm
Nr. layers	20
Growth rate	1.2
s^+, l^+	~313
d^+	3.2

$$d = d_W - f_d \max(0, d_W - C_{\text{DES}} \Delta x)$$

$$f_d = 1 - \tanh[(8r_d)^3] \tag{14.4}$$

$$r_d = \frac{\nu_t + \nu}{\kappa^2 d_W^2},$$

where d_W is the closest distance to the wall, $C_{\text{DES}} = 0.65$ (Shur et al. 1999) and ν_t is described by the SA model. Defined in this way, it can be shown that far away from the wall, the model recovers the Smagorinsky model for ν_t. Inside the wall boundary layer the underlying SA model remains unmodified.

The mesh properties are summarized in Table 14.1. The mesh has been generated using the commercial mesh generator ANSA, which generates hybrid tetrahedral-prism meshes. The mesh uses 34M cells, the lead car surface has been meshed with a surface parallel grid spacing of 3 mm. This equates to a normalized grid spacing of $s^+ = l^+ \sim 313$. This resolution has also been applied to a refinement box around the lead car, which extends some distance into the leeward direction of the train, see Fig. 14.3(top). As can be seen from Fig. 14.3(bottom-left) the bogie cutouts and wheel details are resolved with high accuracy. In wall normal direction 20 prism layers with a growth ratio of 1.2 have been applied to the lead car, the splitter plate and all wheel and bogie details, see Fig. 14.3(bottom-right). This yields a normalized wall normal distance of $d^+ = 3.2$ on the train surface. Hence the mesh is a low Reynolds number mesh and wall functions are not required.

Boundary conditions are summarized in Table 14.2. The inlet velocity has a x-component of $u_x = 29.8\,\text{m/s}$. Based on the characteristic length scale h this gives a Reynolds number of Re =225,000 for viscosity of air ($\nu = 1.58 \cdot 10^{-5}\,\text{m}^2/\text{s}$). For the Spalart-Allmaras based models (URANS and DDES) the inlet turbulent viscosity needs to be specified. We specify $\nu_t = 100\nu$ as is appropriate for internal flows. At the side and outlet patches zero gradient extrapolation is applied for all fields, except that $p = 0$ is prescribed at the outlet patch to avoid pressure drifting. At the wall boundaries (lead car, backcar and plate) we apply standard wall conditions for the velocity and pressure. Note that no wall function is used here for the turbulent

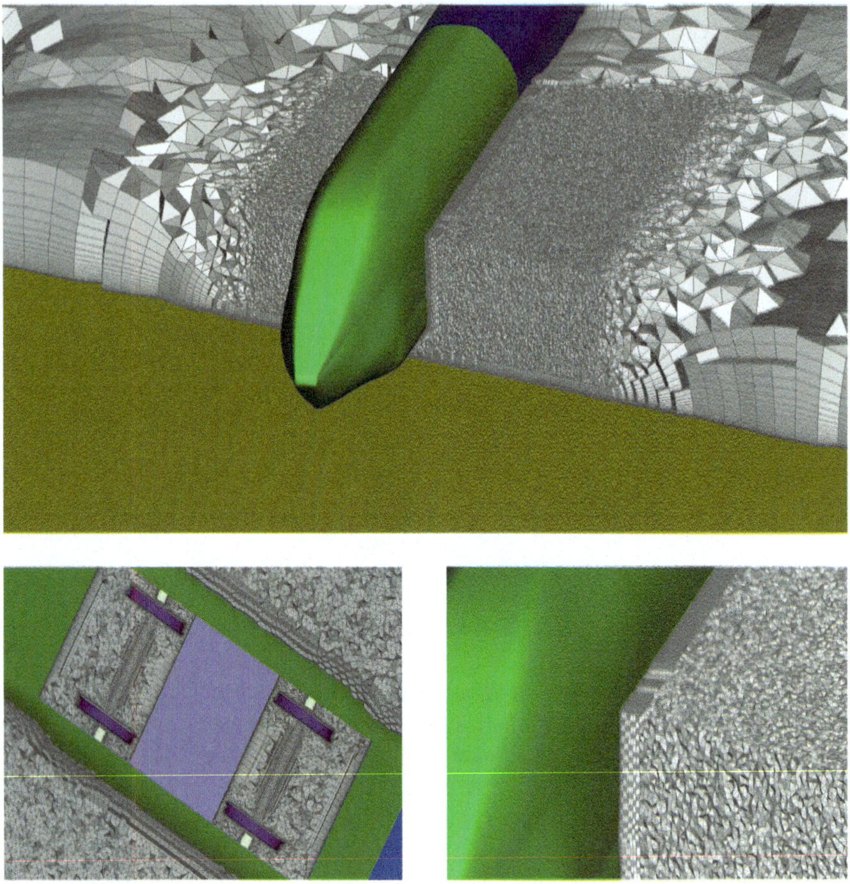

Fig. 14.3 Mesh used in the numerical simulations. *Top* Refinement region around the lead car. *Bottom left* Bogie and wheels section. *Bottom right* Layers applied to the lead car

viscosity. As the low Reynolds number mesh allows resolving the viscous sublayer, the turbulent viscosity should vanish and is accordingly set to zero.

Initial conditions are set in the following way: First velocity and pressure fields are derived from potential theory (i.e. *potentialFoam*). Then a steady RANS simulation based on the standard Spalart-Allmaras model has been performed for $\sim 10,000$ time steps. The resulting fields serve as initial conditions for the URANS, DDES and LES simulations.

The simulation have been run for 50 convective timescales $T_C = h/U_\infty$. After this time force coefficients have reached quasi-steady values with relative standard deviations of 1% in the overturning ($C_{f,y}$) and 2% in the lift coefficient ($C_{f,z}$). The global residual continuity error has converged to $\sim 10^{-13}$ by this time. Running the simulations on a high-performance compute cluster consisting of compute nodes

Table 14.2 Boundary conditions applied at the domain and model geometry patches as indicated by Fig. 14.4

Patch	$\mathbf{u}$[m/s]	p	ν_t
Inlet	$\mathbf{u} = (29.8\ 0\ 0)$	$\nabla p = 0$	$\nu_t = 100\,\nu$
Side	$\nabla u_i = 0$	$\nabla p = 0$	$\nabla \nu_t = 0$
Outlet	$\nabla u_i = 0$	$p = 0$	$\nabla \nu_t = 0$
Frontcar Backcar Plate	$\mathbf{u} = 0$	$\nabla p = 0$	$\nu_t = 0$

Fig. 14.4 Numerical domain with colored boundary conditions. Inlet (*red*), outlet (*magenta* and *turquoise*), model boundaries (*green*, *blue*, *yellow*)

with two 6-core-Intel-Westmere processors each, this takes about 800 h wall time on 80 cores.

14.3 Results

In the following section we will present results of the computations in terms of calculated force coefficients, vorticity slices and total pressure contours. We considered a wind yaw angle of $\beta = 30°$ and Reynolds number of 225,000 for which wind tunnel data is available.

In Table 14.3 we show calculated force coefficients for the URANS, DDES and LES simulations. In the second table the corresponding standard deviations are shown. For the numerical simulations these statistics have been obtained by time averaging over the last 10 convective timescales. The standard deviations are therefore a measure of the unsteadiness in the flow. The experimental standard deviations on the other hand are obtained from ensemble averaging over reproduction measurements. These indicate the experimental error bar associated with the wind-tunnel measurements.

Not shown in the table is the drag force $C_{f,x}$ and corresponding moment $C_{m,y}$. Note that by closing the narrow gap between the lead car and the streamline body in the numerical geometry, drag forces ($C_{f,x}$) differ rather substantially from the experimental data. As is indicated by Table 14.3, the normalized overturning force ($C_{f,y}$) and moment ($C_{m,x}$) are predicted with high accuracy by the URANS and DDES simulations, where relative discrepancies to the wind tunnel data are ~3–4%, which is within the experimental error bar for these components. LES, by contrast, overpredicts this force component significantly, resulting in relative discrepancies to the experimental data of over $\sim 7\%$. This gives a clear indication that the accurate modelling of boundary layer turbulence has a significant influence on the prediction quality of this force component. Noting that $C_{f,y}$ is tightly connected to the flow separation on the leeward/roof interface, which in turn depends strongly on the boundary layer forming prior to separation, such sensitivity is in fact not unexpected.

The prediction of the lift coefficient ($C_{f,z}$) is best for the DDES and LES simulations, while URANS gives the worst results. It should be realized that the lift force is mainly driven by the complex flow in the underfloor region of the train, where interaction with complicated bogie and wheels details takes place. Hence it is expected that a turbulence modelling approach in which the turbulent dynamics on all scales is prescribed by a precalibrated model (URANS) can not suffice to predict this force component accurately. For LES and DDES, by contrast, the largest turbulent eddies forming in the wake flow are resolved rather than modelled. However even for DDES and LES relative deviations of $\sim 10\%$ to the experimental data are evident. It can also be inferred from Table 14.3, that for the simulations the lift force exhibits the largest relative standard deviations ($\sim 2\%$). This is consistent with the observation, that most of the unsteadiness in the flow field is associated with unsteady vortex shedding in the underfloor region.

The twisting moment ($C_{m,z}$) is predicted with high accuracy by LES and DDES, where relative deviations are ~2–4% to experimental data, which is of the same order as the experimental standard deviation ($\sim 3\%$). In contrast URANS results show much larger deviations, i.e. $\sim 8\%$. Since $C_{m,z}$ is negative, this implies that the majority of the aerodynamic loads must accumulate at the front nose of the train, see Fig. 14.2.

To summarize, we find that DDES gives overall the best prediction accuracy of the aerodynamic force coefficients. While LES gives comparable accuracy in the lift force, it overpredicts the overturning force. URANS on the other hand predicts the overturning force accurately, but performs poorly on the lift force.

In Fig. 14.5(left) we display the final snapshot (i.e. at $t = 50T_C$) of the total pressure isosurface along with velocity magnitude for the LES simulation. Total pressure is defined as:

$$p_{tot} = \frac{1}{2}\rho \mathbf{u}^2 + p \tag{14.5}$$

Table 14.3 Force and moment coefficients of the simulations and wind tunnel data. The lower table displays corresponding standard deviations

Simulation	$C_{f,y}$	$C_{f,z}$	$C_{m,x}$	$C_{m,z}$
Wind tunnel	−5.28	2.25	−3.46	−5.54
URANS	−5.44	2.64	−3.60	−5.97
DDES	−5.42	2.55	−3.59	−5.76
LES	−5.67	2.58	−3.81	−5.64
std Wind tunnel	0.22	0.10	0.13	0.16
std URANS	0.00	0.00	0.00	0.00
std DDES	0.04	0.05	0.01	0.05
std LES	0.03	0.04	0.01	0.05

The shown isosurface corresponds to a value for the total pressure which is equal to the free stream pressure, i.e. $p_{tot} = p_\infty = 0$, where the last equation holds because we are solving the incompressible set of equations. Corresponding vorticity slices, which are normal to the x-direction, are displayed on the right hand panel of Fig. 14.5. The component shown corresponds to the projection of the vorticity vector onto the train length direction ($\mathbf{e}_L$) and normalization by the inverse convective timescale, which can be written in compact form by setting $|\mathbf{e}_L| = T_C$. In addition, wall shear stress is depicted on the train surface. With the effective stress tensor $\tau = -2(\nu_t + \nu)\bar{S}$ and surface normal of the train surface elements $\mathbf{A}$, the shown component is computed by:

$$W_i = \tau_{ij} A_j - (A_k \tau_{kl} A_l) A_i \qquad (14.6)$$

which ensures only wall parallel stress is shown. Separation and reattachment lines are indicated by regions of low wall shear stress in Fig. 14.5(right).

The flow may be classified by the following two flow regimes. In the upper part, a relatively steady major vortex system forms on the leeward side, which is responsible for driving the overturning force and moment of the train, i.e. $C_{f,y}$ and $C_{m,x}$ respectively. In the lower part, complicated interaction with the train underfloor, bogie and wheels section of the train, leads to vortices being shed in the stream wise and train length wise directions. This flow is mainly responsible for driving the lift force component and is causing the unsteadiness in the computed forces.

As the flow interacts with the front nose, a counter rotating vortex pair emerges, i.e. vortex V1 and V4 in Fig. 14.5(left). Vortex V1 is born by material that separates over the front section of the nose. As the cross section of the nose grows in the train length direction, streamlines on the windward side of the train will eventually get deflected towards the back of the train and approach the train roof at the end of the nose section. This leads to a slight bifurcation of the flow, i.e. streamlines cluster in the very front and back section of the nose, which causes the formation of two separate vortices, V1 and V2 as seen from Fig. 14.5(left). It is evident that these two vortices merge to form a single vortex at the end of the nose. Once formed, the merged vortex core starts

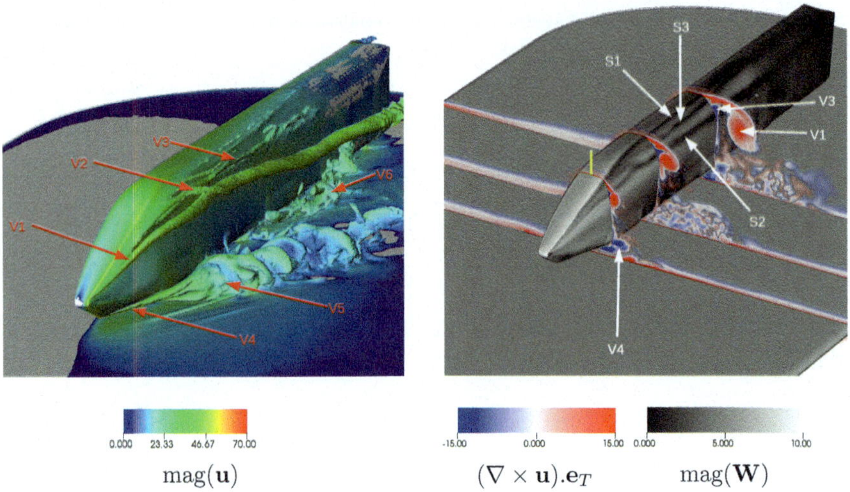

Fig. 14.5 *Left* total pressure contours at $p_{tot} = 0$. Color indicates velocity magnitude. *Right* Vorticity projected on train length direction $\mathbf{e}_T$ on three different slices. The wall shear stress on the train surface mag($\mathbf{W}$) is also depicted. The vertical *yellow lines* indicates the location at which velocity profiles are presented in Fig. 14.7

to deviate from the train surface towards the stream wise direction, possibly because vortex V2, which consists of streamlines that have built up a significant stream wise velocity component, is dominating the merged vortex. These findings are in general agreement with previous studies (Hemida and Krajnović 2009), who found that a number of smaller vortices is born on the leeward/roof interface and merge into the major vortex V1. Flow separation at separation line S1, see Fig. 14.5(right) is feeding the merged vortex which causes it to grow in size, while staying attached to the train surface for almost the entire length of the front car. Attached flow associated with this vortex moves toward the roof of the train and separates on secondary separation line S2. Such secondary separation has been confirmed by previous publications (Fragner et al. 2014) where they were illustrated using surface streamlines. It should be stressed that the formation and merging details of these two vortices (i.e. V1 and V2) have a significant impact on the overturning forces and moments.

The flow streaming over the train nose can further be divided into two vertical regimes. Flow that is located in layers most close to the nose surface, separates and leads to the formation of vortex V1. Flow that is streaming passed the nose in layers above the closest layers, does not separate. Instead this material gets deflected towards the back of the train, where it eventually will form vortex V3 after complex interaction with the vortex merging site of V1 and V2. The resulting vortex V3 is a counter rotating vortex (with respect to V1 and V2) as may be seen from Fig. 14.5(right). Interaction of the counter rotating vortex (V3) with the separating boundary layer (at S1) causes the formation of a small prograde rotating vortex, which causes a region of low wall shear stress at S3.

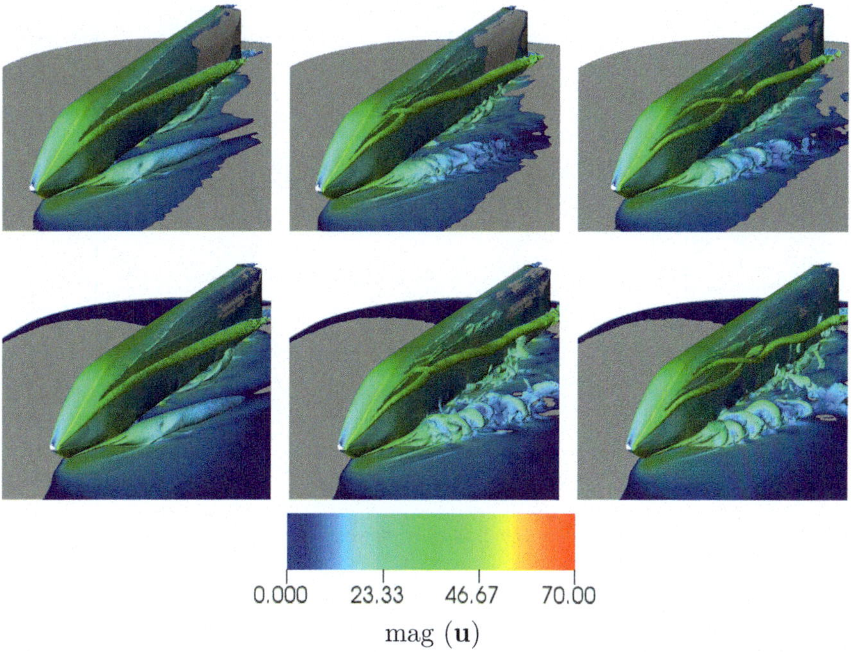

Fig. 14.6 Total pressure contours at $p_{tot} = 0$. Color indicates velocity magnitude. *Left* URANS, *Middle* LES, *Right* DDES. *Top* time averaged profiles, *Bottom* instantaneous snapshot

The vortex system in the bottom region of the train is dominated by unsteady vortex shedding as the flow interacts with the wheels and bogie section details. In the front bogie section, vortices are shed in the stream wise direction (V5). The lower nose vortex (V4), once it is born on the lower part of the train nose, detaches from the surface and is convected downstream in the wake. At the second bogie section flow separation causes vortex shedding in the direction of the train length direction (V6). It should be noted, that most of the unsteadiness in the calculated forces is associated with the vortex shedding dynamics in the underbody flow.

Figure 14.6 shows time averaged and instantaneous total pressure contours for the URANS, LES and DDES simulations. The time averaged plots in Fig. 14.6 show data that has been averaged over the last 10 convective time scales, while the instantaneous plots show the simulations at the final stage, i.e. at $t = 50T_C$.

As may be seen from the Fig. 14.6, there is virtually no difference between the time averaged and instantaneous data for the URANS calculations. These simulations remain very steady without any noticeable unsteady behavior. This is due to the relatively large turbulent dissipation applied in URANS, which is damping out any small-scale unsteady features in the flow. The situation is different for the LES and DDES models. Here the turbulent dissipation in the wake is low enough to allow small-scale unsteady flow features to be present. Owing to the relatively limited resolution of these simulations, however, the unsteadiness is mainly confined to the

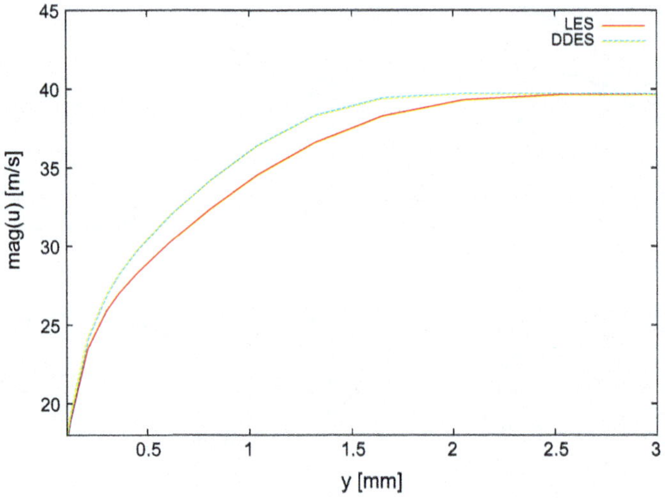

Fig. 14.7 Vertical velocity profiles as indicated by the *yellow line* in Fig. 14.5(right)

vortex shedding in the lower part of the flow. The merged vortex in the upper part remains steady for a distance of a few train heights measured from the front nose, after which it oscillates with low amplitude.

Besides these unsteady flow features there are noticeable differences between the URANS, DDES and LES simulations. First of all, it is clear that vortex V2 is not visible in the URANS simulations. It is likely that the increased turbulent dissipation in URANS prevents bifurcation of the flow on the windward side. Instead of clustering, streamlines are organized rather smoothly over the train nose. In the DDES simulations on the other hand, vortex V2 is clearly visible. However in contrast to LES, vortex V1 bends away from the roof/leeward edge with its core located lower and further downstream within the nose section. When vortex V2 forms, vortex V1 is now at some distance away, which results in the concurrence of both vortices that wrap around each other. The reason that vortex V1 bends away from the leeward/roof edge in the DDES simulation is related to the thickness of the boundary layer forming on the train nose. Figure 14.7 shows the vertical velocity profile for the LES and DDES simulations as indicated by the yellow line in Fig. 14.5(right). It is evident that the boundary layer is smaller in the DDES simulations compared to LES, hence material generally has a larger velocity in DDES. As this material separates to form vortex V1, it is consistent that the core of vortex V1 is further away from the roof/leeward edge in the DDES results. This highlights the fact that the correct modelling of the boundary layer has a significant effect on the formation of the leeward side vortex system. In fact, the turbulent viscosity levels inside the train roof and nose boundary layers are much larger in the LES simulations compared to DDES (not shown), which may indicate that a Smagorinsky model with van Driest damping may not be

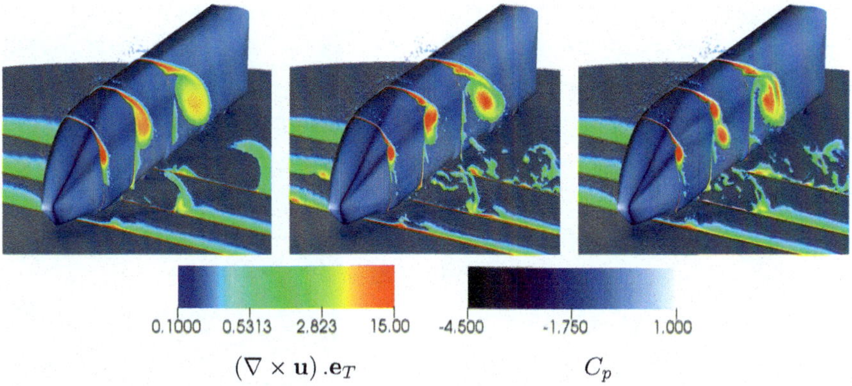

Fig. 14.8 Time averaged vorticity projected on train length direction $\mathbf{e}_T$ on three different slices. Also indicated is the surface pressure coefficient C_P. *Left* URANS, *Middle* LES, *Right* DDES

appropriate for modelling boundary layer turbulence in the presence of severe stream line curvature as induced by the train nose.

Time averaged vorticity is displayed on slices normal to the x-direction in Fig. 14.8 along with the surface pressure coefficient, C_P. Latter is obtained by normalizing the pressure using the dynamic pressure q. As is indicated by Fig. 14.8, the vortex cores associated with the leeward side vortex system are conserved to a much higher quality in the LES and DDES simulations owing to the reduced turbulent dissipation. The double vortex structure in the DDES simulation is illustrated by the two vorticity cores in the middle slice, Fig. 14.8(right).

Footprints of these vortex systems on the train surface are visible from the surface pressure coefficient, C_P. It is clear that the vortices introduce regions of suction pressure on the train leeward side, which act to increase the overturning forces and moments on the train. Comparing LES and DDES, the suction pressure associated with vortex V1 is lower in the DDES results, because in the latter this vortex is at further distance to the train surface. This is the reason that the computed overturning force and moments are higher in the LES results. Comparing LES and URANS on the other hand, vortex V1 is at similar distance to the train surface, therefore introducing similar levels of suction pressure. The difference in the computed overturning force and moment is however driven by the absence of vortex V2 in the URANS results.

Figure 14.9(left) displays the time averaged magnitude of vorticity in the underfloor region of the train, where the time averaging window has been $\Delta t = 10T_C$ as before. On the right hand side of Fig. 14.9 we also depict the fluctuations of vorticity in terms of its variance, where this is computed as the trace of the variance tensor. From the mean vorticity data it is evident, that the interaction with the wheels and bogie details in the bogie-cut out sections leads to very complex flow structures in the bogie cavity and wake region. Regions of high vorticity are associated with flow that has separated after interacting with the wheels or the bogie walls. In addition,

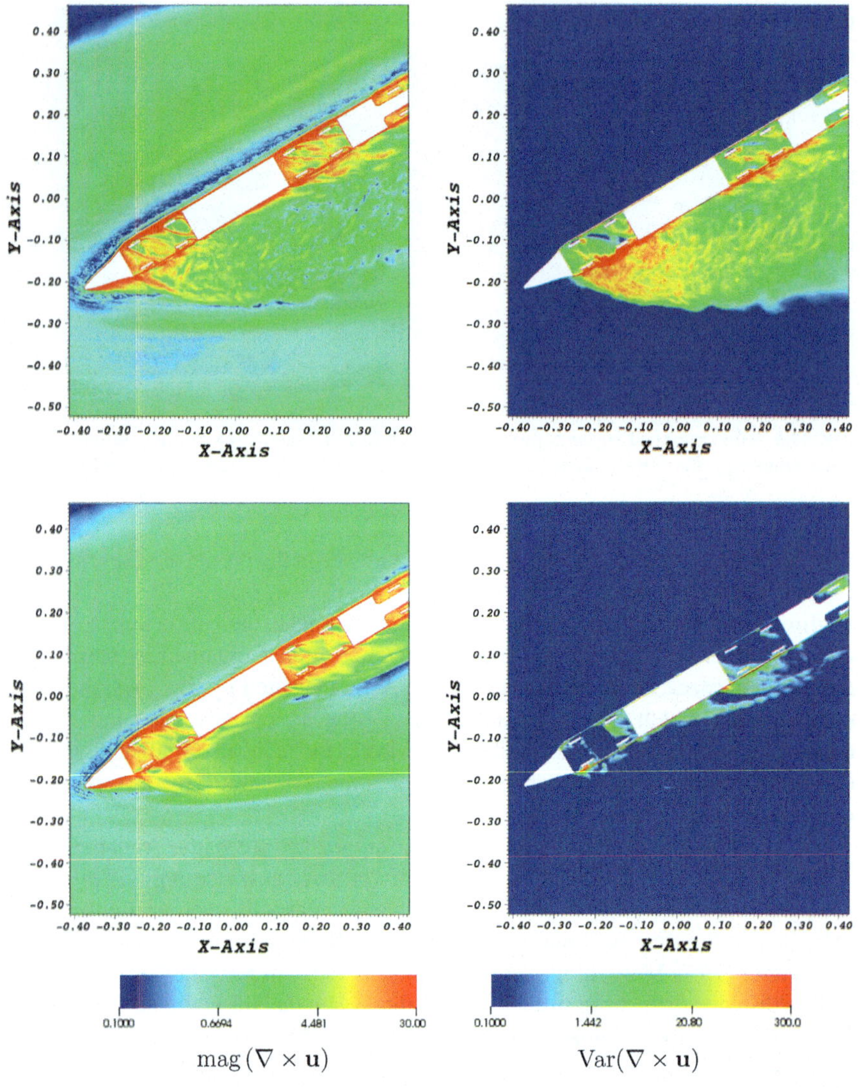

Fig. 14.9 *Left* Mean vorticity in the underfloor region. *Right* Variance of the vorticity. *Top* LES. *Bottom* URANS

the circulating flow inside the bogie cavity is visible from high vorticity in the cavity corners and edges. These structures seem to be much better resolved in LES, while they appear diffused in the URANS simulation. More importantly, it may be seen from Fig. 14.9(right) that time fluctuations are significant in the LES simulations, while URANS shows almost no time fluctuations by comparison. In LES the shown variance equates to relative standard deviations of up to ~50–60% of the vorticity magnitude in the near wake flow. Further downstream, the time fluctuations start to decay.

It is not expected that the effect of these highly unsteady small-scale turbulent fluctuations on mean flow quantities can be properly represented by a pre-calibrated RANS model. Hence it is not surprising, that we observe a relative poor prediction quality in the URANS simulations with respect to the lift coefficient, latter which is mostly affected by the underbody flow. DDES and LES, by contrast, deliver improved results. The resolved turbulent kinetic energy is given as $E_R = \frac{1}{2}Var(\mathbf{u})$, while the modelled turbulent kinetic energy is $E_M = (\nu_t/L)^2$, where L is the turbulent length scale. Assuming $L = 1$ mm which corresponds to the smallest cell sizes in the bogie cavity region, we confirm that the ratio $E_R/(E_M + E_R)$ is significantly above 80%. Hence most of the turbulent kinetic energy is resolved in the DDES and LES simulations, explaining the improved prediction accuracy with respect to the lift coefficient for these simulations.

14.4 Conclusion

In this paper we have investigated the influence of various turbulence modeling approaches on crosswind flow simulations, where we compared unsteady RANS, delayed detached eddy (DDES) and large eddy simulations (LES). The model we employed is the DLR's Next Generation Train 2 (NGT2) geometry, and we consider the crosswind flow at a Reynolds number of Re =225,000 and yaw angle of $\beta = 30°$. Calculated forces and moments are contrasted to wind tunnel measurements. The main results of this paper can be summarized as follows:

- The flow may be classified by the following two flow regimes. In the upper part, a relatively steady major vortex system forms on the leeward side. In the lower part, complicated interaction with the train underfloor, bogie and wheels section of the train, leads to vortices being shed in the stream wise and train length wise directions.
- The details of the leeward side vortex system have a large influence on the overturning forces and moments on the train, as suction pressures are induced on the train leeward side.
- The formation of this vortex system depends strongly on the turbulence modelling technique employed. The corresponding overturning forces and moments are over-predicted by LES, while they are quite accurately predicted by URANS and DDES ($\leq$3% deviation to experimental data). This indicates that the correct modelling of the train roof boundary layer, i.e. via an appropriate RANS model, is important to improve the prediction of this force component.
- The lift coefficient is strongly affected by the dynamics of the underbody flow.
- Unsteady vortex shedding in the train underfloor region is mainly responsible for the time fluctuations in all calculated forces.

- Small-scale unsteady turbulent eddies associated with this region are resolved in the LES and DDES simulations. This proved to be crucial to deliver improved prediction results for the lift coefficient.
- Because of its ability to use a RANS model inside the train roof boundary layer on the one hand, and to resolve small-scale turbulent eddies in the underfloor region on the other hand, we find, that DDES gives the best prediction quality of the force and moment coefficients.

The last conclusion should have general applicability to crosswind flow about complex geometries at industry relevant Reynolds numbers, where resolution requirements for wall resolving LES are not affordable, yet complicated geometric features generate turbulent wake flow that cannot be described by a precalibrated RANS model. The results in this paper should be verified using a high resolution mesh. We are currently running simulations using a mesh employing up to 220M cells for LES and DDES. Although these simulations are somewhat far away from final convergence, first results seem to indicate, that higher resolution is particularly beneficial for predicting the lift force as was already indicated in this paper. The high resolution results will be presented in a subsequent publication.

References

Baker CJ (2003) Some complex applications of the wind loading chain. J Wind Eng Ind Aerodyn 91(12):1791–1811

Baker CJ, Jones J, Lopez-Calleja F, Munday J (2004) Measurements of the cross wind forces on trains. J Wind Eng Ind Aerodyn 92(7):547–563

Bürkle D, Winter J (2013) Forschen für den Zug der Zukunft. http://www.dlr.de/dlr/en/desktopdefault.aspx/tabid-10467/740_read-916/gallery/2043/740_read-916/

Chiu TW, Squire LC (1992) An experimental study of the flow over a train in a crosswind at large yaw angles up to 90. J Wind Eng Ind Aerodyn 45(1):47–74

Copley JM (1987) The three-dimensional flow around railway trains. J Wind Eng Ind Aerodyn 26(1):21–52

Diedrichs B (2003) On computational fluid dynamics modelling of crosswind effects for high-speed rolling stock. Proc Inst Mech Eng Part F J Rail Rapid Transit 217(3):203–226

Ferziger JH, Peric M (2012) Computational methods for fluid dynamics. Springer Science & Business Media, Berlin

Ferziger JH, Peric M, Leonard A (1997) Computational methods for fluid dynamics. Phys Today 50:80

Fragner M, Weinman K, Deiterding R, Fey U, Wagner C (2014) Numerical and experimental studies of train geometries subject to cross winds. Civil-Comp Press, Stirlingshire, UK

Haff J, Richard J, Fey U, Kowalski T, Loose S, Wagner C (2012) Wind tunnel experiments with a high-speed train model subject to cross wind conditions. In: Proceedings of the first international conference on railway technology: research, development and maintenance. Civil-Comp Press, Stirlingshire, Scotland

Hemida H, Krajnović S (2008) LES study of the influence of a train-nose shape on the flow structures under cross-wind conditions. J Fluids Eng 130(9):091101

Hemida H, Krajnović S (2009) Exploring flow structures around a simplified ICE2 train subjected to a 30 side wind using LES. Eng Appl Comput Fluid Mech 3(1):28–41

Hemida H, Baker C (2010) Large-eddy simulation of the flow around a freight wagon subjected to a crosswind. Comput Fluids 39(10):1944–1956

Hemida H, Krajnović S (2010) LES study of the influence of the nose shape and yaw angles on flow structures around trains. J Wind Eng Ind Aerodyn 98(1):34–46

Jasak H (1996) Error analysis and estimation for finite volume method with applications to fluid flow. Ph.D. thesis, Imperial College London

Khier W, Breuer M, Durst F (2000) Flow structure around trains under side wind conditions: a numerical study. Comput Fluids 29(2):179–195

Krajnović S (2014) On unsteady simulations in train aerodynamics. In: Railways 2014 the second international conference on railway technology: research, development and maintenance Ajaccio, Corsica, France 8–11 April 2014

Krajnović S, Ringqvist P, Nakade K, Basara B (2012) Large eddy simulation of the flow around a simplified train moving through a crosswind flow. J Wind Eng Ind Aerodyn 110:86–99

Morden JA, Hemida H, Baker CJ (2015) Comparison of RANS and detached Eddy simulation results to wind-tunnel data for the surface pressures upon a class 43 high-speed train. J Fluids Eng 137(4):041108

Östh J, Krajnović S (2014) A study of the aerodynamics of a generic container freight wagon using large-eddy simulation. J Fluids Struct 44:31–51

Shur M, Spalart PR, Strelets M, Travin A (1999) Detached-eddy simulation of an airfoil at high angle of attack. Eng Turbul Model Exp 4:669–678

Spalart PR, Jou WH, Strelets M, Allmaras SR et al (1997) Comments on the feasibility of LES for wings, and on a hybrid RANS/LES approach. Adv DNS/LES 1:4–8

Spalart PR, Deck S, Shur ML, Squires KD, Strelets MK, Travin A (2006) A new version of detached-eddy simulation, resistant to ambiguous grid densities. Theor Comput Fluid Dyn 20(3):181–195

Weller HG, Tabor G, Jasak H, Fureby C (1998) A tensorial approach to computational continuum mechanics using object-oriented techniques. Comput Phys 12(6):620–631

Chapter 15
Russian Mechanism to Support Renewable Energy Investments: Before and After Analysis

Mariia Kozlova, Mikael Collan and Pasi Luukka

Abstract This chapter presents an analysis of how the new Russian support policy for renewable energy investments changes the expected profitability of renewable energy investments in Russia. A comparative analysis of investment profitability in the before and after support policy cases is presented for a wind farm investment to illustrate the effect of the policy. This chapter is among the first to comparatively analyze the effect of the Russian renewable energy support mechanism on investment project profitability.

15.1 Introduction

To ensure strategic investments in emerging technologies in the energy sector such as investments in renewable energy (RE) that cannot at this time compete with conventional solutions in industrial scale and in terms of profitability, policy makers may introduce support mechanisms. Design of support mechanisms has turned out to be a crucial element in being able to incentivize the deployment of RE technologies into the energy markets. This is due to the maturity level of the present day technology—RE power projects are typically not profitable without policy support. It remains to be seen how fast the development of technology is able to change the situation. Ideally, a supporting policy for RE should serve multiple goals, such as reducing the risks of investment and providing motivation to invest cost efficiently, it should also be feasible from the point of the society that is paying for the bill. Typically, the

M. Kozlova (✉)
School of Business and Management, Lappeenranta University of Technology, BOX 20, 53851 Lappeenranta, Finland
e-mail: mariia.kozlova@lut.fi

M. Collan · P. Luukka
Lappeenranta University of Technology, BOX 20, 53851 Lappeenranta, Finland
e-mail: mikael.collan@lut.fi

P. Luukka
e-mail: pasi.luukka@lut.fi

P. Diez et al. (eds.), *Computational Methods and Models for Transport*,
Computational Methods in Applied Sciences 45,
DOI 10.1007/978-3-319-54490-8_15

main RE support policy types used include a "remuneration scheme" that guarantees an investor supplemental income to strengthen profitability, if an investment fulfills the policy given criteria. The level of the supplemental income can be defined in different ways and common ways to do this include, e.g., using a special price to be paid for electricity produced by RE investments (feed-in tariffs); exploring the minimum acceptable electricity price through competitive auctions; and letting the markets define the price by establishment of RE certificate trade, which RE producers receive for their production (Boute 2015).

In this chapter we concentrate on studying the Russian RE support mechanism for the Russian wholesale energy market. Under the present day conditions, industrial-scale renewable energy production investments in Russia are seldom considered as a relevant option without the support mechanism and would be deemed "universally" unprofitable without one. The focus of this chapter is to study the effect the Russian RE incentive mechanism has on the profitability of Russian RE investments. The analysis presented is based on the codified details of the Russian RE mechanism and its pricing instructions that available in the original Russian language legislative procedures (Russian law) (Government of Russian Federation 2013a, b, 2015), in a policy report (International Finance Corporation 2013), and a recently published by academic paper (Kozlova and Collan 2016). Likely due to the lack of English language sources on the Russian RE policy it has received relatively little attention in the academic literature.

Previous literature that studies the Russian RE policy includes a qualitative study of the draft version of policy (Boute 2012), an analysis of its impact on the Russian electricity and capacity prices (Vasileva et al. 2015), and a number of investigations concerning RE investment profitability in Russia (Kozlova and Collan 2015, 2016; Kozlova et al. 2015). In this respect the recent paper (Kozlova and Collan 2016) is the closest match to the analysis presented in this chapter and studies the effects of different factors on RE project profitability under the Russian capacity-based support and examines the policy's interim success. The study is limited to using classical investment analysis and sensitivity analysis and in those terms offers a simplified picture. The aim of this research is to perform a comparative "before – after" study on the effect of the Russian RE policy to RE investments in Russia. To the best of our knowledge this is a first time the results of such a study are reported.

As a basis for the study, we use a previously presented (Kozlova 2015) system dynamic investment model and use simulation analysis to study the profitability of a wind farm investment in Russia with and without the supporting policy. The results show that the policy has a significant effect to project profitability, when a project fulfills the policy goals.

This chapter continues by shortly presenting the Russian support mechanism for renewable energy, then a simulation analysis of the profitability of a stylized wind farm investment case is made, ceteris paribus, with and without the supporting policy in place, and finally the paper is closed with a discussion and conclusions are drawn.

15.2 Russian Capacity Market and the Support Mechanism for RE Investments

Russian capacity mechanism for renewable energy support is an extension to the pre-existing Russian national capacity trade principles. The Russian power market consists of a capacity market that operates alongside the electricity market (Gore et al. 2012; Government of Russian Federation 2010a). The idea behind capacity trade is to assure ability of a power system to meet electricity demand in the long term by timely incentivizing investments in new power plants. Selling capacity means getting paid for being available to produce electricity (Olsina et al. 2014).

The Russian capacity market is organized through competitive capacity selection that is carried out by the centralized infrastructure organization System Operator (SO), where existing power generators submit their bids with available volumes of installed capacity for a pre-specified period (NP Market Council 2012). In addition, new planned projects compete for long-term capacity delivery agreements with regulated price (Boute 2012). Capacity power generators that have won contracts are obliged to follow dispatching orders from the SO that enable the management of the power system operation. On the demand side, each electricity buyer in the wholesale market is obliged to buy capacity according to the buyer's peak demand. The price of capacity is defined as a weighted average of contracted and auctioned capacity prices that include the "winning bids" of auctions and the regulated tariffs of long-term capacity delivery agreements (Gore and Viljainen 2014). Notably, the procedure of capacity price calculation for the long-term agreements is designed to assure a riskless return on each investment project (Government of Russian Federation 2010b). This mechanism is taken as a foundation for designing the remuneration scheme for renewable energy support. In May 2013, the Government of Russian Federation presented an extended capacity mechanism that is specific to RE power generation, with the aim to support the deployment of almost 6 GW of new renewable energy capacity by the year 2020 (Government of Russian Federation 2013a, b). Once a year, competitive capacity auctions are conducted for investment projects into wind, solar PV, and small (<25 MW) hydropower. The selection of projects is carried out for a four-years-ahead commercialization window and is based on (i) compliance to participation requirements and on (ii) the least planned capital costs criterion. There is a set yearly target installed capacity volume for each particular RE technology, up to which projects are selected. The participation requirements include a technology-specific capital expenditure limit and a requirement to procure a share of the used equipment from local Russian manufacturers (local content) (International Finance Corporation 2013).

The selected projects are eligible for a long-term capacity delivery agreement that allows them to benefit from monthly capacity payments for fifteen years, starting from these projects' commercialization date. The capacity agreement comes into force only after a qualification procedure after the construction end in accordance with (Government of Russian Federation 2008). The procedure registers a power plant as one operating on a renewable energy source and controls and confirms the fulfillment of the local content requirement. The policy obliges the winning RE projects to start operations on time and penalizes for delays, while the obligation that in place for "conventional energy production" facilities to follow dispatching orders is "softened" to the requirement of complying with SO orders to switch off electricity production (Government of Russian Federation 2013a).

The RE capacity price calculation is centralized and the procedure is designed to guarantee a specific return on investment regardless of the changing market conditions. The guaranteed return is defined as 12% annually corrected by changes in market interest rates (or 15% for projects auctioned before 1.01.2016). The market interest rates are represented by long-term Russian government bond yield. The capacity payments are designed to cover project costs and to provide some return over it. The estimated project costs comprise of capital expenditures (CapEx), operating expenses (OpEx), and of taxes. The capital expenditures are directly taken from the submitted bid information and are project-specific. They are converted into monthly payments by means of variable rate annuity. The 'foreign' share of the capital costs is translated to rubles during the project investment phase. For operating expenses the mechanism sets technology-specific norms that are corrected with inflation, which is incorporated into the calculation by using the consumer price index as a proxy. Project-specific property tax forecast, based on planned capital costs and a 20% income tax are included in the estimated project expenses for capacity price calculation.

The complex calculation of the capacity payments introduces a complicated "effect" that influences project profitability analysis. The key market and project-internal factors directly affect project profitability, while they also enter in the capacity price payment computations and create "cross-effects" on project profitability (see Fig. 15.1). It can be understood that the procedure is rather complex. For details we refer the interested reader to study Appendix 2 of (Kozlova and Collan 2016).

Capacity sales are not the only source of revenues for a power plant project, there is also income from the electricity sales. This is why the capacity payments do not cover all the estimated project costs, but only the share of costs that is computed via

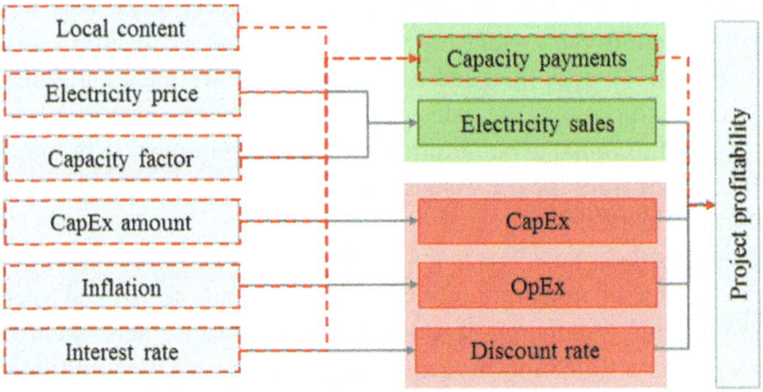

Fig. 15.1 The factors affecting Russian RE investment profitability

an analysis of an averaged project. This share of expenses is also a subject to annual recalculation, based on changing electricity prices and changes in other conditions (Government of Russian Federation 2013a, 2015).

The mechanism uses several ways to shape investor motivation to invest in RE projects, typically non-conformity with requirements decreases the support received, such cases include situations where, the plant qualification procedure reveals that the local content requirement is not fulfilled, the volume of the yearly produced electricity is too low, and orders of the System Operator to switch off the electricity production to maintain the balance in the system are not followed.

To summarize, the RE capacity price is calculated for each project on an annual basis and the aim is to provide a riskless return on the RE investment. The capacity price is adjusted based on the changing market conditions that include interest rates, electricity prices, inflation, and exchange rates. Project-specific factors such as capital costs, local content, and electricity production performance are taken into account. More detailed information about the policy and the capacity pricing can be found in (Boute 2012; International Finance Corporation 2013; Kozlova and Collan 2016) and from the original Russian legislation (Government of Russian Federation 2013a, b, 2015).

In the following section of this chapter we present a case and a numerical illustration that uncovers the effect of the Russian RE support mechanism on the profitability of a wind farm investment.

15.3 Case: Effect of the Russian RE Support Mechanism on a Wind Farm Investment Profitability

The analysis presented is based on the use of a system dynamic model built to study the profitability of a stylized wind farm investment. The model used has been presented in detail in (Kozlova 2015). The model realized with Matlab Simulink®, is built to fully represent the details of the Russian RE support mechanism, and the profitability calculation part of the model is based on using a typical discounted cash-flow logic that returns the project net present value (NPV) as a result. The model is used as the basis for a Monte Carlo simulation (Hacura et al. 2001; Kwak and Ingall 2007). The simulated NPV results are presented as histograms. We have chosen a 10 MW wind farm as the case to be studied, solar PV and small hydropower projects generate very similar results. The wind farm is assumed to be commissioned in 2017, and to start generating cash-flows immediately thereafter for the next 20 years. In the base case investment the total capital costs are assumed to be equal to cost level set by the Russian legislation limit of 110 Mrub./MW. The operating costs are assumed to be equal to the normative 188 Krub./MW per month, adjusted with inflation. We assume inflation to be an uncertain variable and assume it to stay within a range of 1 to 1.7, in terms of consumer price index. The revenues are treated as uncertain and they are modeled to consist of electricity sales with uncertain price that are assumed

Table 15.1 Summary of the parameter values and settings for the simulation runs

Run	Policy	CapEx (of the limit 110 Mrub./MW)	Capacity factor (from target 27%)	Local content	Electricity price	Consumer price index
A	*Not in place*	*Uncertain 100–150%*	*Uncertain 30–100%*	–	*Uncertain 1–3 rub./kWh*	*Uncertain 1–1.7*
B	In place	*Uncertain 100–150%*	*Uncertain 30–100%*	*Uncertain*	*Uncertain 1–3 rub./kWh*	*Uncertain 1–1.7*
C	In place	Certain 100%	High 90–100%	Fulfilled	*Uncertain 1–3 rub./kWh*	*Uncertain 1–1.7*

to range from 1 to 3 rub./kWh (corresponding to typical prices on the markets), by a capacity factor that is equal to the target, and the capacity payments that are calculated according to the support mechanism procedure. Values of the uncertain variables are expected to have a uniform distribution, except for the "local content" variable that is binary and may only take the value zero or one. The Russian risk-free rate used in all calculations is assumed to be fixed at ten percent—its effects are studied separately.

Three simulation runs are performed with the model, one without the support policy (run A), one with the support policy (run B), and a third one to illustrate the situation, where a project is able to completely fulfill the requirements of the support policy and thus being able to enjoy the full benefits of the said policy (run C). Each simulation run consisted of 100,000 simulation rounds. Details for the three simulation runs are visible in Table 15.1.

Market uncertainty represented in the model by electricity prices and inflation is modeled equally through the all three runs. The run A represents no policy situation, thus the project receives no capacity payments and benefits only from electricity sales. We assume possibility of CapEx increase to a maximum of 150% of the basic value of 110 Mrub./MW. The capacity factor used in the RE support mechanism depends on resource availability and we assume it to vary within a broad range from 30 to 100% of the target capacity factor that is equal to 27% for wind power.

The difference in run A without the supporting policy in place and run B, where the policy is in place, is the appearance of the supporting subsidy payments and the "local content" variable. The local content is typically fulfilled, as the investors most likely will not start project construction if the local content requirement cannot be fulfilled, because this would mean almost certainly that the project is unprofitable.

Run C illustrates the situation when the project is able to comply with all policy requirements and achieves the full support payments without any penalties for under-performance. Figure 15.2 presents the resulting NPV distributions from the three runs as histograms.

We can see from Fig. 15.2 how the introduction of the supporting mechanism changes the NPV distribution of the investment (Fig. 15.2a, b). What is important to notice is that the distribution resulting from the "no support mechanism" case

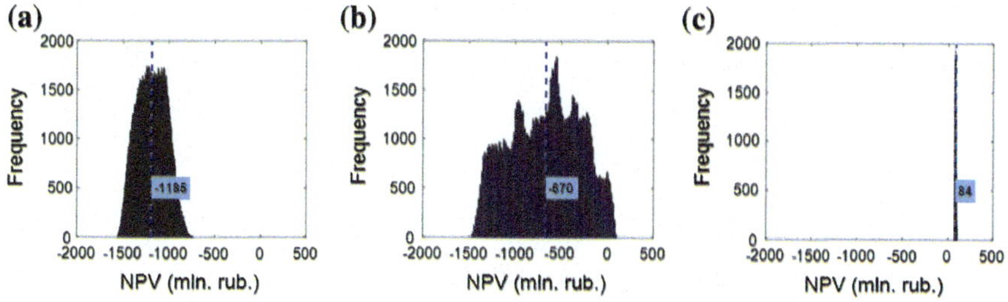

Fig. 15.2 NPV distributions for three scenarios (expected mean highlighted)

(run A) consists of only negative profitability outcomes, while the presence of the supporting mechanism shifts the distribution to the right and creates a possibility for profitable investment.

Results from run A are realistic in the sense that industrial-scale RE investments have not been profitable in Russia without support. The distribution that results from run B has a multi-peak shape that is caused by complex rules behind the support mechanism that determines the capacity remuneration paid to the investment under different circumstances. The lower bound of the distribution remains the same in both cases, because in case of poor project performance also a project "under" the support mechanism remains without any support. The difference between the two distributions can be simply calculated as the difference between the mean NPV values of runs A and B, which amounts to 515 million rubles for the investment project or more generally 52 million rubles per 1 MW of wind power installed capacity. This can also be interpreted as a real option value that is generated by the support mechanism. Comparing the means may not be very useful while both are negative, therefore we have also separately studied the situation of a project that is able to fulfill all the requirements of the support policy (run C).

The resulting distribution from run C is very concentrated and entirely in the positive area. This means that under the studied conditions a project that is able to fulfill all the set requirements of the support policy is able to guarantee positive NPV with a small variation of outcomes even in the presence of the uncertain market factors. This highlights the fact that it makes sense for the investors to be proactive in pushing their performance to fulfill the requirements of the policy. This is also in line with the policy objectives. For such projects the value of the support policy, in comparison to the preceding situation without the policy, is quite remarkable and stands at 1269 million rubles, or 127 million rubles per MW. This can be considered a real option value.

The Russian RE capacity mechanism shields investors not only from the price risk, but also from the interest rate risk. Therefore, the effects of changing interest rates on the project profitability require separate attention. The capacity price is computed as an annuity with variable rate that is adjusted to the changes in the local risk-free rate. Thus, an increase in the risk-free rate leads to higher capacity payments, and consequently to a higher internal rate of return of the project. Technically speaking,

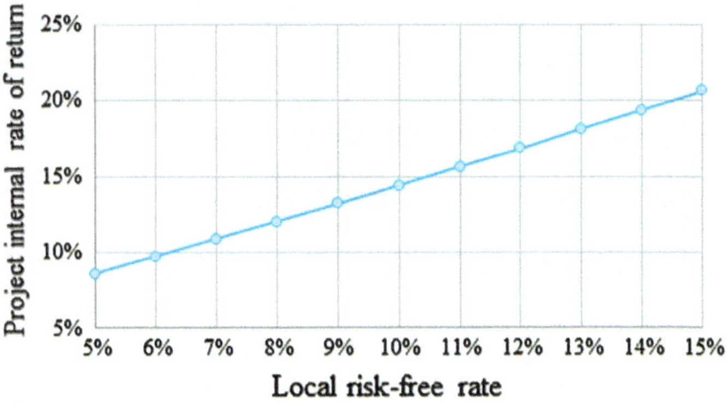

Fig. 15.3 Sensitivity of project internal rate of return (IRR) to the Russian risk-free rate under the support policy

without such mechanism in place the higher interest rates would reduce the NPV due to higher discount rates. The design of Russian RE support offsets this reduction effect by offering an increased capacity price when interest rates rise and aims to keep project profitability on the same level. This feature is designed to enable investors to cope with higher costs of financing without harming the project profitability.

A numerical sensitivity analysis performed of the effect of the Russian risk-free rate to the project profitability shows that there is a close to linear positive relationship between the project IRR and the risk-free rate arising from the remuneration design, see Fig. 15.3.

This is a non-insignificant analysis in the Russian context, where the risk-free rate may experience considerable changes over time. What is interesting and important to note is that the choice of contract in terms of floating or fixed interest on the project debt plays a role in project profitability: change in the interest rate of a floating rate loan is most likely offset by the support mechanism to a large extent.

In general, the results show that the Russian RE support mechanism is able to provide higher profitability to RE projects and projects that are able to consistently fulfill the set requirements may enjoy a situation of low risk profitability. This also means that investors should proactively try to steer their investments into meeting the set criteria.

15.4 Discussion and Conclusions

In this chapter we showed how the Russian renewable energy support mechanism works and how it changes the profitability outlook of RE power investments in Russia. The Russian RE support mechanism is unique in terms of its construct, which is rather complex. The mechanism aims to guarantee profitability for investments that fulfill the set criteria by offering capacity payments to RE investments. The capacity pricing mechanism that determines the size of the payments starts by defining the required return on investment and then uses it to compute a remuneration amount needed for a particular project to achieve this return. This remuneration is a subject to yearly recalculations in that take into account the changing market environment and the project performance in the long-term. As the result, the mechanism allows investors a lower risk with regards to the project profitability and allows them to enjoy a greater level of independence from Russian market conditions if they comply with set policy requirements. Real world results from the implementation of the said policy so far suggest that the support mechanism is able to provide sufficient incentive for investors for investment in solar power. Experience with wind and small hydropower shows that the selection of parameter values for the support mechanism needs more work with regards to these technologies.

We demonstrated with a system dynamic model of a wind power generation investment how the Russian RE support mechanism changes the profitability landscape of the investment—there is a significant and an important effect that allows projects that fulfill the set policy requirements to be profitable. This chapter is among the first, if not the first, to comparatively analyze the effect of the Russian support mechanism on the profitability of renewable energy investments in Russia.

Further research into this topic will include comparing the results with results from using other commonly used RE support mechanisms.

Acknowledgements The authors would like to acknowledge the support received by M. Kozlova from the Fortum Foundation (Grant no.201500068).

References

Boute A (2015) Annual reporting of renewables: ten years of excellence, renewable energy policy network for the 21st century

Boute A (2012) Promoting renewable energy through capacity markets: an analysis of the Russian support scheme. Energy Policy 46:68–77

Gore O, Viljainen S (2014) Challenges of cross-border trade between two markets with different designs. In: 11th international conference on the european energy market (EEM14). IEEE

Gore O, Viljainen S, Makkonen M, Kuleshov D (2012) Russian electricity market reform: deregulation or re-regulation? Energy Policy 41:676–685

Government of Russian Federation (2008) Decree #426 on the qualification of renewable energy sources installations

Government of Russian Federation (2010a) Decree #1172 approving the wholesale market rules

Government of Russian Federation (2010b) Decree #238 on the determination of price parameters for the trade in capacity

Government of Russian Federation (2013a) Decree #449 on the mechanism of promoting the use of renewable energy in the wholesale market of electric energy and power

Government of Russian Federation (2013b) Resolution #861-R on amendments being made to resolution #1-R 8.01.2009 on the main directions for the state policy to improve the energy efficiency of the electricity sector on the basis of renewable energy sources for the period up to 2020

Government of Russian Federation (2015) Decree #1210 on the introduction of amendments to the certain legislative acts regarding the use of renewable energy sources in the wholesale electricity and capacity market

Hacura A, Jadamus-Hacura M, Kocot A (2001) Risk analysis in investment appraisal based on the Monte Carlo simulation technique. Eur Phys J B-Condens Matter Complex Syst 20(4):551–553

International Finance Corporation (2013) Russians new capacity-based renewable energy support scheme, an analysis of Decree no. 449

Kozlova M (2015) Analyzing the effects of the new renewable energy policy in Russia on investments into wind, solar and small hydro power. M.Phil. thesis, Lappeenranta University of Technology

Kozlova M, Collan M (2015) Financing renewable energy projects in Russia under new legislation

Kozlova M, Collan M (2016) Modeling the effects of the new Russian capacity mechanism on renewable energy investments. Energy Policy 95:350–360

Kozlova M, Collan M, Luukka P (2015) Renewable energy in emerging economies: shortly analyzing the russian incentive mechanisms for renewable energy investments. In: International research conference GSOM emerging markets conference-2015: business and government perspectives

Kwak YH, Ingall L (2007) Exploring Monte Carlo simulation applications for project management. Risk Manag 9(1):44–57

NP Market Council (2012) Wholesale market

Olsina F, Pringles R, Larisson C, Garcés F (2014) Reliability payments to generation capacity in electricity markets. Energy Policy 73:211–224

Vasileva E, Viljainen S, Sulamaa P, Kuleshov D (2015) RES support in Russia: impact on capacity and electricity market prices. Renew Energy 76:82–90

The manufacturer's authorised representative in the EU is Springer Nature Customer Service Centre GmbH, Europaplatz 3, 69115 Heidelberg, Germany. If you have any concerns regarding our products, please contact ProductSafety@springernature.com

Printed and bound by CPI Group (UK) Ltd, Croydon, CR0 4YY
30/09/2026
02225501-0002